GENETIC ENTROPY & The Mystery of the GENOME

– the genome is degenerating.

Dr. J.C. Sanford

GENETIC ENTROPY
&
The Mystery of the
GENOME

Second Edition

ISBN 1-59919-002-8

Published by
Elim Publishing
Lima, New York

About the Author

Dr. John Sanford has been a Cornell University Professor for more than 25 years (being semi-retired since 1998). He received his Ph.D. from the University of Wisconsin in the area of plant breeding and plant genetics. While a professor at Cornell he trained graduate students and conducted genetic research at the New York State Agricultural Experiment Station in Geneva, NY. During this time John bred new crop varieties using conventional breeding, and became heavily involved in the newly emerging field of plant genetic engineering. While at Cornell, John published over 70 scientific publications, and was granted over 25 patents. His most significant scientific contributions involved three inventions - the biolistic ("gene gun") process, pathogen-derived resistance, and genetic immunization. Most of the transgenic crops grown in the world today were genetically engineered using the gene gun technology developed by John and his collaborators. John also started two successful biotech businesses deriving from his research - Biolistics, Inc. and Sanford Scientific, Inc. John still holds a position at Cornell (Courtesy Associate Professor), but has largely retired from Cornell and has started a small non-profit organization - Feed My Sheep Foundation.

Dedication and Acknowledgements

I feel I could only write this little book by God's grace, and acknowledge and thank Him as the giver of every good thing. This book is dedicated to the memory of Dr. Bob Hanneman, my professor and graduate thesis advisor, who encouraged me in my science, and provided an example for me regarding faith and godliness. I would like to thank my wife, Helen, for her unswerving support; and my friend, Fred Meiners, for his encouragement. Special thanks to Walter ReMine for his computer simulation of mutation accumulation, and Lloyd R. Hight for his artwork. I acknowledge Michael Behe and many others who went before me in recognizing that the Primary Axiom is not true. I am the last, and the least, to recognize this.

Genetic Entropy

& The Mystery of the GENOME

Foreword

By Dr. John Baumgardner

During the past half century, the scientific enterprise has opened a door into an almost surrealistic Lilliputian realm of self-replicating robotic manufacturing plants, with components whirring at tens of thousands of RPM, with automated parcel addressing, transport, and distribution systems, with complex monitoring and feedback control. Of course, this is the realm of cell and molecular biology. It is a realm in which tens of thousands of different kinds of sophisticated nanomachines perform incredible chemical feats inside the living cell. Above and beyond this cellular complexity is the equally complex realm of the organism, with trillions of cells working in astonishing coordination, and above that is the realm of the brain - with its multiplied trillions of neural connections. Confronted with such staggering complexity, the reflective person naturally asks, "How did all this come to exist?" The standard answer given to this question is what the author of this book calls "the Primary Axiom."

Genetic Entropy and the Mystery of the Genome represents a probing analysis of the fundamental underpinnings of the Primary Axiom. In particular, it focuses on the genetic software that specifies life's astounding complexity. The author points out that for higher organisms, and certainly for humans, the extent of these genetic specifications, called the genome, is vast. Not only is the genome huge, it is also exceedingly complex. It is filled with loops and branches, with genes that regulate other genes that regulate still other genes. In many cases, the same string of genetic letters can code for entirely different messages–depending on its context. How such an astonishing information structure has come into existence is clearly an important question. But the author introduces a further question, namely, how can the human genome even be *maintained* against the degrading effects of billions of new deleterious mutations, which enter the human population each generation?

Concerning the Primary Axiom, the author acknowledges that as a professional geneticist he, for many years, discerned no serious problems with its theoretical underpinnings. He confides that during his training in graduate school he accepted this primarily by trust in the authorities, rather than by genuine personal understanding. At that point he felt he had no choice—he thought this abstract and highly mathematical field was beyond his own ability to assess critically. It was not until much later in his professional career that he became aware of how unrealistic and how vulnerable to critical analysis were the crucial assumptions on which the Axiom rests. The author concludes that most professional biologists today are just like he was earlier in his career. Most simply are not aware of the fundamental problems with the Axiom. This is because the Axiom's foundational assumptions are not critiqued in any serious way, either in graduate classes, or in graduate level textbooks, or even in the professional literature.

The conceptual models that population genetics has offered to the rest of the professional biology community, presented in the guise of mathematical elegance, have at their foundations a number of unjustifiable assumptions. The Primary Axiom, it turns out, depends on these assumptions for its support. Most professional biologists are simply not aware of this state of affairs.

The field of population gentics deals largely with complex mathematical models that attempt to describe how mutations, after they arise, are passed from one generation to the next and how they affect the survival of individual members of a population in each generation. The reality of these conceptual models depends critically, of course, upon the realism of the assumptions on which they are built. In this book the author exposes the obvious lack of realism of many of the most crucial assumptions that have been applied for the past 75 years. Most professional biologists, like the author during the earlier part of his professional career, base much of their confidence in the Primary Axiom on claims derived from these conceptual models that have employed observationally unjustifiable assumptions. Most biologists today are unaware that the population genetics claims to which they were exposed in graduate school can no longer be defended from a scientific standpoint. Most, therefore, can hardly imagine that when realistic assumptions are applied, population genetics actually repudiates the Axiom.

The Mystery of the Genome is a brilliant expose on the un-reality of the Primary Axiom. It is written in a challenging but accessible style, understandable by non-specialists with a modest background in either genetics or biology. At the same time, this book has sufficient substance and documentation to cause the most highly trained biologist to rethink in a serious way what he or she probably has always believed about the Primary Axiom. In my opinion, this book deserves to be read by every professional biologist and biology teacher in the world. To me it has the potential of changing the outlook of the academic world in a profound way.

John Baumgardner has a Ph.D. in geophysics from UCLA and worked as a research scientist in the Theoretical Division of Los Alamos National Laboratory for 20 years. He also received a M.S. degree in electrical engineering from Princeton University, where he first became aware of information theory - and later its implications for biological systems. He is an expert in complex numerical simulations.

Prologue

In retrospect, I realize that I have wasted so much of my life arguing about things that don't really matter. It is my sincere hope that this book can actually address something that really does matter. The issues of *who we are, where we come from, and where we are going* seem to me to be of enormous importance. This is the real subject of this book.

Modern thinking centers around the premise that man is just the product of a pointless natural process - undirected evolution. This very widely taught doctrine, when taken to its logical conclusion, leads us to believe that we are just meaningless "bags of molecules", and in the last analysis - nothing matters. If false, this doctrine has been the most insidious and destructive thought system ever devised by man (yet if it is true, it is at best like everything else - meaningless). The whole thought system which prevails within today's intelligentsia is built upon the ideological foundation of undirected and pointless Darwinian evolution.

Modern Darwinism is built, most fundamentally, upon what I will be calling "The Primary Axiom". The Primary Axiom is that man is merely the product of *random mutations* plus *natural selection*. Within our society's academia, the Primary Axiom is universally taught, and almost universally accepted. It is the constantly-mouthed mantra, repeated endlessly on every college campus. It is very difficult to find any professor on any college campus who would even consider (or should I say - dare) to question the Primary Axiom. It is for this reason that the overwhelming majority of youth who start out with a belief that there is more to life than mere chemistry - will lose that faith while at college. I believe this is also the cause of the widespread self-destructive and self-denigrating behaviors we see throughout our culture.

What if the Primary Axiom were wrong? If the Primary Axiom could be shown to be wrong it would profoundly affect our culture; and I believe it would profoundly affect millions of individual lives. It could change the very way we think about ourselves.

Late in my career, I did something which for a Cornell professor would seem unthinkable. I began to question the Primary Axiom. I did this with great fear and trepidation. By doing this, I knew I would be at odds with the most "sacred cow" within modern academia. Among other things, it might even result in my *expulsion* from the academic world. Although I had achieved considerable success and notoriety within my own particular specialty (applied genetics), it would mean I would have to be stepping out of the safety of my own little niche. I would have to begin to explore some very big things, including aspects of theoretical genetics which I had always accepted by faith alone. I felt compelled to do all this – but I must confess that I fully expected to simply hit a brick wall. To my own amazement, I gradually realized that the seemingly "great and unassailable fortress" which has been built up around the Primary Axiom is really a house of cards. The Primary Axiom is actually an extremely vulnerable theory – in fact it is essentially indefensible. Its apparent invincibility derives largely from bluster, smoke, and mirrors. A large part of what keeps the Axiom standing is an almost mystical faith, which the *true-believers* have in the omnipotence of natural selection. Furthermore, I began to see that this deep-seated faith in natural selection is typically coupled with a degree of ideological commitment - which can only be described as religious. I started to realize (again with trepidation), that I might be offending a lot of people's religion!

To question the Primary Axiom required me to re-examine virtually everything I thought I knew about genetics. This was the most difficult intellectual endeavor of my life. Deeply entrenched thought patterns only change very slowly (and I must add - painfully). What I eventually experienced was a complete over-throw of my previous understandings. Several years of personal struggle resulted in a new understanding, and a very strong conviction that the Primary Axiom was most definitely wrong. More importantly, I became convinced that the Axiom could be ***shown*** to be wrong to any reasonable and open-minded individual. This realization was exhilarating, but again - frightening. I realized that I had a moral obligation to openly challenge this most sacred of cows. In doing this, I realized I would earn for myself the intense disdain of most of my colleagues within academia - not to mention very intense opposition and anger from other high places.

What should I do? It has become my conviction that the Primary Axiom is insidious on the highest level – having catastrophic impact on countless human lives. Furthermore, every form of objective analysis I have performed has convinced me that the Axiom is clearly false. So now, regardless of the consequences, I have to say it out loud: ***the Emperor has no clothes***!

I invite the reader to carefully consider this very important issue. Are you really just a meaningless bag of molecules - the product of nothing more than random molecular mutations and reproductive filtering? As you read this book, I am going to ask you to wrap your mind around something very challenging but also very exciting. I contend that if you will invest a reasonable mental effort as needed to follow just a handful of fairly simple arguments - I can persuade you that the Primary Axiom is false. Can you imagine anything more radical, or more liberating? To the extent that the Primary Axiom can be shown to be false, it should have a major impact on your own life - and on the world at large. For this reason, I have dared to write this humble little book – which some will receive as blasphemous treason, and others - revelation.

If the Primary Axiom is wrong, then there is a surprising and very practical consequence. When subjected only to natural forces, the human genome must irrevocably degenerate over time. Such a sober realization should have more then just intellectual or historical significance. It should rightfully cause us to personally reconsider where we should rationally be placing our hope for the future.

The genome is the book of life. Where did it come from?

Newsflash - The genome is an instruction manual.

The **genome*** is the instruction manual which specifies life. The human genome is the instruction manual which specifies human cells to be human cells, and specifies the human body to be the human body. There is no information system designed by man that can even begin to compare to it.

The complex nature of the genome can only be appreciated when we begin to grasp how much information it contains. When you assemble the little red wagon for your child on Christmas Eve, there is a booklet that tells you how to put it together. The size of the booklet is deceptive - it does not contain all the information needed for fabricating the component parts, nor manufacturing the steel, rubber, and paint. The complete instruction manual would actually be a very substantial volume. If you compiled all the instruction manuals associated with creating a modern automobile, it would comprise a substantial library. That library would be very large if it included the information needed for making all the components, as well as all the information needed for the robotic assembly lines.

* *An organism's genome is the sum total of all its genetic parts, including all its chromosomes, genes, and nucleotides.*

Likewise, the manuals required for creating a jet fighter and all its components, computers, and assembly lines, would comprise an extremely large library indeed. The manuals needed for building the entire space shuttle and all its components and all its support systems would be truly enormous! Yet the *specified complexity* of even the simplest form of life - a bacterium - is arguably as great as that of the space shuttle. Now in this light - try to absorb the fact that the jump in complexity from a bacterium up to a human being is arguably as great as the jump from the little red wagon up to the space shuttle! There is simply no human technology that can even begin to serve as an adequate analogy for the complexity of a human life. Yet the genome is the instruction manual encoding all that information - as needed for life!

We have thus far only discovered the first dimension of this "book of life" – which is a linear sequence of 4 types of extremely small molecules, called nucleotides. These small molecules make up the individual "steps" of the spiral-staircase structure of DNA. These molecules are the *letters* of the genetic code, and are shown symbolically as A, T, C, and G. These letters are strung together like a linear text. They are not just symbolically shown as letters, they *are* very literally the *letters* of our instruction manual. Small clusters or motifs of these four molecular letters make up the *words* of our manual, which combine to form genes (the *chapters* of our manual), which combine to form chromosomes (the *volumes* of our manual), which combine to form the whole genome (the entire *library*).

A complete human genome consists of two sets of 3 billion individual 'letters' each. Only a very small fraction of this genetic library is required to directly encode the roughly 100,000 different human proteins, and the uncounted number of functional human

RNA molecules which are found within our cells. Each of these protein and RNA molecules are essentially miniature *machines*, each with hundreds of component parts, each with its own exquisite complexity, design, and function. But the genome's *linear* information, equivalent to many complete sets of a large encyclopedia, is not enough to explain the complexity of life.

As marvelous as all this linear information is, it must only be the first dimension of complexity within the genome. The genome is not just a simple string of letters spelling out a linear series of instructions. It actually embodies multiple linear codes, which overlap and constitute an exceedingly sophisticated information system, embodying what is called "data compression" (Chapter 9).

In addition to multiple, overlapping, linear, language-like forms of genetic information, the genome is full of countless loops and branches - like a computer program. It has genes that regulate genes that regulate genes. It has genes that sense changes in the environment, and then instruct other genes to react by setting in motion complex cascades of events that can then modify the environment. Some genes actively rearrange themselves, or modify and methylate other gene sequences - basically *changing* portions of the instruction manual!

Lastly, there is good evidence that linear DNA can fold into two- and three-dimensional structures (as do proteins and RNAs), and that such folding probably encodes still higher levels of information. Within the typical non-dividing nucleus, there is reason to believe there may be fabulously complex three-dimensional arrays of DNA, whose 3-D architecture controls higher biological functions.

The bottom line is this: the genome's set of instructions is not a simple, static, linear array of letters; but is dynamic, self-regulating, and multi-dimensional. There is no human information system that can even begin to compare to it. The genome's highest levels of complexity and interaction are probably beyond the reach of our understanding. Yet we can at least acknowledge that this higher level of information must exist. So while the linear information within the human genome is limited to approximately the contents of numerous complete sets of encyclopedias, the non-linear information must obviously be much, much greater. Given the unsurpassed complexity of life, this necessarily has to be true.

All this mind-boggling information is contained within a genomic package that is contained within a cell's nucleus - a space much smaller than the smallest speck of dust. Each human body contains a "galaxy" of cells (more than 100 trillion), and every one of these cells has a complete set of instructions and its own highly prescribed duties. The human genome not only specifies the complexity of our cells and our bodies, but also the functioning of our brains - a still higher level of organization which is entirely beyond our comprehension.

As we recognize the higher-order dimensions of the genome, I believe we can readily agree with Carl Sagan (1974) that each cell contains more information than the Library of Congress. Indeed, surely human life is more complex than all human technologies combined! **Where did all this information come from, and how can it possibly be maintained? This is the mystery of the genome.**

The standard answer to the origin of biological information is that *mutation* combined with *selection* have created all biological information. All genomes (manuals) must have derived from some simple "first" genome - via a long series of mutations (typographical errors) and lots of natural selection (differential copying). In fact, this is the *Primary Axiom* of biological evolution. **Life is life because random mutations at the molecular level are filtered through a reproductive sieve on the level of the whole organism.** But just what is an axiom? An axiom is an untestable concept, which is accepted by faith because it seems so obviously true to all reasonable parties. On this basis, it is accepted as an Absolute Truth. In this book, I am going to urge the reader to ask the question: "Should we accept today's Primary Axiom?" If the Primary Axiom could be shown to be wrong, it would mean that our current understanding of the history of life is also wrong. This would justify a *paradigm shift** of the highest magnitude, and would allow us to completely re-evaluate many of the deeply entrenched concepts which frame modern thinking.

It is important that we put the Primary Axiom into a framework that is both honest and is also realistic to our mind's eye. I would like to propose an honest analogy which very accurately characterizes today's Primary Axiom. My analogy involves the evolution of transportation technologies, as outlined below.

The first primeval genome encoded the assembly instructions for the first "little red wagon". That simple genomic instruction manual was copied by an invisible mechanical 'scribe', to make more instruction manuals. Each newly copied manual was used to make a new red wagon. However, the scribe, being imperfect,

**a paradigm shift is a change in some very fundamental idea, which has governed a group's collective thinking.*

made errors. So each wagon came out different. Each wagon had its own unique instruction manual taped to its bottom. When the first wagons were junked, their instruction manuals were also junked. New copies of instruction manuals could only be made from the manuals of the immediately preceding generation of wagons, just before they were to be discarded. Since the copying of instructions was sequential (rather than going back to an original master copy), errors accumulated over time in all manuals, and the resulting wagons started to change and vary. The accumulating errors are, of course, our analogy for mutations.

Perhaps you are uneasy with this picture? No doubt you realize that we are looking at a deteriorating situation. Information is being lost, instructions are becoming muddled or degraded, and the wagons will doubtless deteriorate in quality. Eventually, the system will break down, the manual will become complete gibberish, and workable wagons will become extinct. We will examine this problem of mutation in more detail in Chapters 2 and 3.

At this point we introduce our hero - natural selection. Natural selection is like a judge, or quality control agent, deciding which wagons are suitable models for further copying. Natural selection, as the judge, instructs the scribe not to copy manuals from "inferior wagons". This represents differential reproduction (reproductive sieving) - which is selection. But it is important to understand there is never direct selection for "good instructions" - only for "good wagons". As we will see, this is very important. Mutations are complex and happen at the molecular level, but selection can only be carried out on the level of the whole organism. The scribe and judge work entirely independently. The scribe is essentially blind, working on the level of molecules; and being extremely nearsighted, he can only see that individual letter which he is copying. The judge is also nearly blind, but he is extremely far-sighted - he never sees the letters of the manual, or even the wagon's individual components. He can only see the relative performance of the whole wagon.

The scribe can be envisioned at the beginning of a robotic assembly line; he copies programs for the robots by blindly and imperfectly duplicating older programs, just one binary bit at a time. The quality control agent looks at the performance of the finished wagons, and decides which wagons are better than others. The programs from the wagons which he has chosen are then given to the scribe - for the next round of copying and assembly.

The story continues – many defective wagons can be eliminated, and so presumably most errors in the instructions might likewise be eliminated. More exciting - some rare spelling errors might result in *better* wagons, and so the judge can instruct the scribe to preferentially copy these instructions. The process of evolution has begun! We will be examining the feasibility of the selection process as a mechanism for improving genomic information in this way.

The story goes on - the information within the instruction manual might not only be *improved* by this process, but it can also be *expanded*. If the imperfect scribe occasionally copies an extra (duplicate) page out of the manual, we might start <u>adding</u> information. Naturally, a duplicate page in an instruction manual is not really new information - in fact it will invariably confuse and disrupt the reading of the manual. But again, the judge only allows copying of manuals from *good wagons*. So, bad duplications might presumably be eliminated, and harmless duplications might be preserved. Now these harmless duplications will also begin to have copying errors within them, and some of these errors *might* create new and useful information - new instructions for new functional components in the wagon. With a little imagination, perhaps we can picture a variety of duplications eventually evolving, via misspellings, and specifying something entirely new - like an internal combustion engine, or wings, or an on-board computer navigational system. Hence we have a scenario whereby a little red wagon can, through a series of typographical errors within its manual, evolve into an automobile, a plane, or even the space shuttle.

Yet this analogy still does not go far enough, because a human being is much more complex than a space shuttle. In fact, our **phenome** (our entire body including our brain), is immeasurably more complex than any human technology. Perhaps we can come closer to the mark, if we imagine our little red wagon being transformed into the fanciful *Star Ship Phenome* - complete with "warp speed engines" and a "holodeck" (Figure 1). This is really the best analogy. The primary Axiom says that misspellings and some differential copying can simultaneously explain the library (the genome), and the starship (the phenome), which are illustrated in Figure 1d.

We must now ask: could misspellings and selective copying really do this? A correct understanding of *selection* is essential for evaluating the merit of the Primary Axiom. No intelligence is involved in this scenario. The scribe is really just a complex array of senseless molecular machines that blindly replicate DNA. The judge is just the tendency for some things to reproduce more than other things. Many people unconsciously attribute to natural selection a type of supernatural intelligence. But actually, natural selection is just a term for a blind and purposeless process whereby some things reproduce more than others. It is crucial we understand that our scribe and judge have neither foresight nor intelligence - their combined I.Q. equals *zero*.

Isn't it remarkable that the Primary Axiom of biological evolution essentially claims that typographical errors plus some selective copying can transform a wagon into a spaceship, in the absence of any intelligence, purpose, or design? Do you find this concept credible? It becomes even more startling when we realize that under the Primary Axiom, the spaceship was in no way pre-specified - not even in the mind of God. It truly "just happened" - by accident. The spaceship is essentially just a *mutant wagon*. Yet this illustration is actually the best analogy for describing the Primary Axiom. The only weakness of this analogy is that there is no human technology that can compare to the actual complexity of life, and thus there is no human information system that can compare to the human genome.

This whole analogy stands in sharp contrast to the false picture portrayed by Dawkins (1986). The famous Dawkins argument, built around the phrase "me thinks it is like a weasel", involved a *pre-specified message*, which is systematically uncovered through a simple-minded process equivalent to the children's games "20 Questions" or "Hangman". In Dawkins' model, both the message and the carefully-crafted and finely tuned method of uncovering it, are intelligently designed and purposeful. Furthermore, his selection scheme allows for direct selection of genotype (misspellings) rather than phenotype (wagon performance).

Briefly, Dawkins set up a simple computer program which started with a simple random array of letters, having exactly the same number of characters as the phrase "me thinks it is like a weasel". He designed his program to then begin to randomly "mutate" the letters. When a new letter fell into place which matched the message "me thinks it is like a weasel" the program would select that "improved" message. Obviously it would not take long for such a little program to create the desired phrase. However, even to make this simple program work, Dawkins had to carefully design the "replication rate" and "mutation rate" and other parameters to get the results he wanted. This program supposedly proved that evolution via mutation/selection is inevitable - not requiring any intelligent design. Obviously, Dawkins used an intelligently designed computer, and then he used his own intelligence to design the program, optimize it, and even to design the pre-selected phrase. Dawkins' argument cannot honestly be applied (for many reasons) to defend the Primary Axiom – which does not allow for the operation of any intelligence, purpose, or forethought - and does not allow for direct selection for misspellings themselves.

In this book, we are going to examine some basic aspects of genetics and determine if the known facts about the human genome are compatible with the Primary Axiom. As you read this book, if you come to the point where you feel that the Primary Axiom is no longer <u>obviously true to all reasonable parties</u>, then rationally at that point, you should feel obligated to reject it as an *axiom*. If the Primary Axiom cannot stand up as an axiom (should anything?), then it should be treated as an unproven hypothesis - subject to falsification.

Figure 1a.
Some assembly required...

A little red wagon is not information, but it requires information to specify its assembly. A typical assembly booklet is not really all the information required to specify the production of a little red wagon. The truly complete production manual would be a very substantial book - specifying the production of all the components (i.e. wheels, etc.), and all raw materials (steel, paint, rubber).

Figure 1b.

The complete instructions needed to specify a modern automobile would comprise a very substantial library. If the assembly was to be done entirely by machines (no "intelligence" required), the information, including that required for making and programming the robots, would be massive, comprising a very large library.

Figure 1c.

The complete instruction manual needed to specify a jet fighter, including its on-board computer systems and all the manufacturing and support systems inherent in creating and maintaining such a system, would be a massive library. Imagine the instructions, if every component had to be made robotically!

Figure 1d.

The library shown above represents the human genome (all our genetic information). The spaceship represents the human phenome (our entire body, including our brain). We cannot really imagine how extensive a library would have to be if it were to specify the fictional S.S. Phenome - complete with warp-speed engines and a holodeck. Wouldn't it have to be much larger than the Library of Congress? Yet it can be reasonably argued that a human is still more complex than a hypothetical S.S. Phenome (what type of starship could reproduce itself?).

Are random mutations good?

Newsflash – Random mutations consistently destroy information.

The subject of mutation in the human genome should be approached with sensitivity, because people matter - and people are hurt by mutation. The number of families affected by birth defects is tragically high - so this is not just a matter of "statistics". Genetic disease, in its broadest sense, is catastrophic. If we include all genetic predispositions to all pathologies, we must conclude that we are *all* highly "mutant". Furthermore, nearly every family is impacted by the tragedy of cancer - which is fundamentally the result of mutations within our body cells. Indeed, growing evidence indicates that *aging* itself is due to the accumulation of mutations within our body cells. Mutations are the source of immeasurable heartache - in fact they are inexorably killing each one of us. So mutations are more than just an academic concern!

Can we say mutations are good? Nearly all health policies are aimed at reducing or minimizing mutation. Most personal health regimes are aimed at reducing mutations, to reduce risk of cancer and other degenerative diseases. How can anyone see mutation as good? Yet according to the Primary Axiom, mutations are good because they create the variation and diversity which allows selection and evolution to occur, creating the information needed for life.

Before we go further, we need to realize there are two types of variation - random variation and designed variation. Random variation is the type of variation I see in my car as time passes - it is the rust, the dings, the scratches, and broken parts. Such things create variation in cars - but do they ever lead to better cars? Can typographical errors realistically improve a student's term paper? Can throwing rocks improve a glass house? Apart from accidents, there exists another type of variation - designed variation. When I bought my car I had many options - paint color, type of tire, size of engine, etc. These options were useful to me in making my selection. These designed variable components have also proven useful to me later on - I have added or taken away various options, replaced broken parts, etc. These designed forms of variation are beneficial, being useful for sustaining my car - and are even useful for *improving* my car - within limits (such variations, even when they are intelligently designed, will never transform my car into a spaceship).

Part of the Primary Axiom is that all genetic variation *must* come from random mutations, since no genetic variation by design is allowed. However, now that the era of genetic engineering has begun, this axiomatic assumption clearly is not true. Many living organisms now contain genetic variations designed and engineered by man. Therefore, the Axiom that all genetic variation comes from random mutation is clearly not true today. Perhaps this simple fact can open our minds to the possibility of designed genetic variation which preceded man! Apart from our ideological commitment to the Primary Axiom, it can very reasonably be argued that random mutations are never good. Speaking in terms of vehicles - they appear to be the *dings* and *scratches* of life, rather than the spare parts.

The overwhelmingly deleterious nature of mutations can be seen by the incredible scarcity of clear cases of information-creating mutations. It must be understood that scientists have a very sensitive and extensive network for detecting information-creating mutations – most scientists are diligently keeping their eyes open for them all the time. This has been true for about 100 years. The sensitivity of this observational network is such that even if only one mutation out of a million really unambiguously creates new information (apart from fine-tuning), the literature should be absolutely over-flowing with reports of this. Yet I am still not convinced there is a single, crystal-clear example of a known mutation which unambiguously *created* information. There are certainly many mutations which have been described as "beneficial", but most of these beneficial mutations have not created information, but rather have destroyed it. For illustration, some of us (like me) would view a broken car alarm as "beneficial". However, such random changes, although they might be found to be "desirable", still represent a breakdown - not the creation of a new functional feature. Information decreases. This is the actual case, for example, in chromosomal mutations for antibiotic resistances in bacteria, where cell functions are routinely lost. The resistant bacterium has not evolved – in fact it has digressed genetically and is *defective*. Such a mutant strain is rapidly replaced by "superior" (i.e. natural) bacteria as soon as the antibiotic is removed. Another example would be the hairless Chihuahua dog. In extreme heat, reduction of size and loss of hair may be useful adaptation, but this involves degeneration, not creation of new information. In such situations, although local adaptation is occurring, information is actually being lost - not added. Yet the Primary Axiom still insists that mutations are good, and are the building blocks with which evolution creates the galaxy of information which currently exists within the genome. Let us continue to examine this concept more closely.

The nature of genetic deterioration via mutation can easily be seen using our analogy of an instruction manual. For example, a single line within a jet aircraft assembly manual might read as follows:

> ***Step 6. When you have completed the last step, go back and repeat step 3, until part B is 10.004 mm thick. Then wait, no less than 3h, before going to the next step.***

Limiting ourselves to just simple point mutations (misspellings) - there are three possible levels of impact on the above instructions. Theoretically, some misspellings might have zero impact on the message (I don't see any obvious examples in this instance). Most misspellings will have a very subtle effect on the clarity or coherence of the message (i.e. misspellings within all the portions not underlined). Lastly, a few changes (within the underlined areas), will have the potential for dramatic (essentially lethal) effects. What is *not* clear in the above example is which misspellings could actually improve the instructions to result in a better jet plane. While such changes are *conceivable*, they are unlikely - on a level that is difficult to fully describe. Any *possible* improvement in the instructions deriving from a misspelling would be expected to be very slight (i.e. changing the 4 to an 8, three places to the right of the decimal in the specified thickness). These types of changes do not actually *add* information, but really only *modulate,* or fine-tune, the system. It should be obvious to any reasonable person that we can't expect *any* misspellings that would result in a major advance in jet technology. For example, no misspelling in the above sentence is going to create a new patentable component. Such major changes would obviously require intelligent design. For every hypothetical misspelling that might very subtly improve (more accurately, modulate) a jet plane's blueprint, there would be a multitude of misspellings which would be detrimental. The detrimental changes would range from a few lethal errors to a very large number of nearly-neutral changes in the text.

The above illustration can be extended to the genome (see Figure 2). There are over 3 billion potential point mutation sites in the human genome. Only a small fraction of these, when mutated, will have a major effect. Yet none of the potential mutations can be conclusively shown to have zero effect. The vast bulk of the nucleotide positions need to be considered as being "nearly-neutral" sites (as will be seen more clearly below). Misspellings in life's instruction manual will sometimes be very deleterious, but in the overwhelming majority of cases they will be only very slightly deleterious. No new information can be expected, although existing information can be modulated or fine-tuned to a limited extent. Biological modulation would involve adjusting the cell's "rheostats". For example, it is well known that mutations can adjust activity of a promoter or enzyme, either up or down. However, when we use a rheostat to dim a light, we are not creating a new circuit, nor are we in any way creating new information. We are just fine-tuning the system that is already there – which was, in fact, *designed to be fine-tuned.*

I have just stated that the overwhelming majority of mutations should be nearly-neutral. All population geneticists would agree. Why is this? It can be seen for many reasons. Firstly, it can be seen by the nature of misspellings in any written language (as you can picture for yourself by changing any single letter in this book). Secondly, it can be seen by the total number of nucleotides. On average, each nucleotide position can only contain one 3-billionth of the total information. Thirdly, it can be seen from innumerable studies on the mutation of specific coding sequences, promoters and enhancers. Experimentally, we can show that most nucleotide positions have very subtle effects on any given cell function - and only a few mutations are real "killers" of gene function (and remember - any single gene function is just a miniscule part of the whole cell's system). Lastly, the near-neutral impact of most nucleotides can be seen from the very subtle role that single nucleotides play in genome-wide patterns (codon-preferences, nucleosome binding sites, isochores, 'word' compositional differences between species, etc.). These patterns involve hundreds of millions of nucleotides which are dispersed throughout the genome. Individual nucleotide positions must play an immeasurably tiny role in maintaining all such genome-wide patterns. Yet as infinitesimal as these effects are, they are not zero - such patterns exist because each nucleotide contributes to it – each nucleotide still has an impact, and so carries information. No matter how we analyze it, we will see that overwhelmingly, most nucleotide positions must be nearly-neutral.

Are there *truly neutral* nucleotide positions? True neutrality can never actually be demonstrated experimentally (it would require infinite sensitivity). However, for reasons we will get into later, some geneticists have been eager to minimize the functional genome, and have wanted to relegate the vast bulk of the genome to "junk DNA" . So mutations in such DNA would be assumed to be entirely neutral. However, actual research findings relentlessly keep expanding the size of the functional genome, while the presumed "junk DNA" keeps shrinking. In just a few years, many geneticists have shifted from believing that less than 3% of the total genome is functional, to believing that more than 30% is functional - and that fraction is still growing. As the functional genome expands, the likelihood of neutral mutations shrinks. Moreover, there are strong theoretical reasons for believing that there is no truly neutral nucleotide position. By its very existence, a nucleotide position takes up space, affects spacing between other sites, and affects such things as regional nucleotide composition, DNA folding, and nucleosome binding. If a nucleotide carries *absolutely zero* information, it is then by definition slightly deleterious – as it slows cell replication and wastes energy. Just as there are really no truly neutral letters in an encyclopedia, there are probably no truly neutral nucleotide sites in the genome. Therefore there is no way to change any given site, without *some* biological effect - no matter how subtle. Therefore, while most sites are probably "nearly-neutral", very few, if any, should be absolutely neutral.

So what does the real distribution of all mutations really look like? Figure 3a shows the naive view of mutation - a bell-shaped curve, with half of the mutations being beneficial and half being deleterious. It is easy to envision "selective progress" with such a distribution of mutations. Selection would obviously favor the good mutations and eliminate the bad. In fact if this distribution was correct, progressive evolution would be inevitable - a real "no-brainer". In reality, this view is clearly incorrect. BENEFICIAL MUTATIONS ARE SO RARE THAT THEY ARE TYPICALLY NEVER EVEN SHOWN IN SUCH GRAPHS. Figure 3b shows a more realistic view of the distribution of mutations, ranging from lethal (-1) to neutral (0). However, this is still not quite right. Mutations are sharply skewed toward neutral values (in other words most mutations are nearly-neutral, as we have just discussed). What does the real distribution of mutations look like? Figure 3c is modified and expanded from Kimura (1979). This curve very nearly represents the true distribution of mutations.

As can be seen from Kimura's curve, most mutations are negative, and pile up steeply near the zero mark - as we have been saying. In other words they are deleterious and overwhelmingly nearly-neutral. Kimura is famous for showing that there is a "zone of near-neutrality" (shown here as a box). Kimura calls near-neutral mutations "effectively neutral" – meaning that they are so subtle that they are *not subject to selection*. However, we can see that Kimura does not show *any* mutations as being *absolutely* neutral – his curve approaches, but does not reach, the zero-impact point. Kimura's somewhat arbitrary cut-off point for "un-selectable" (i.e. the actual size of his box) he calculates as a function of "N_e" – the number of reproducing individuals within a breeding population.

It is important to note that Kimura's box size, which he calculates based upon population size, is only a minimal estimate of the extent of the effectively neutral mutations. The actual box size should also be enlarged by any and all **non-genetic factors** that can affect reproductive probability. As we will see in Chapter 6, this fact very *significantly increases the size of the box* (see Figure 9). *Anything that decreases the "signal-to-noise ratio" will make proportionately more of a genome's nucleotides utterly un-selectable.* The importance of non-genetic factors in terms of making proportionately more nucleotides un-selectable is acknowledged by the famous geneticist Muller (Muller, 1964).

In Kimura's figure, he does not show any mutations to the right of zero – i.e. there are zero beneficial mutations shown. He obviously considered beneficial mutations so rare as to be outside of consideration. Given this distribution of mutations, one would naturally ask, "How can theorists possibly explain evolutionary progress?" It is done as follows: everything in the "near-neutral box" is redefined as being *completely neutral*, and is thereby dismissed. It is then assumed that the mutations to the left of the near-neutral box can be entirely eliminated using natural selection. Having eliminated *all* deleterious mutations in these two ways the theorists are then free to argue that no matter how rare beneficial mutations may be (to the right of the box), there should now be enough time and enough selection-power left over to rescue them and to use them as the building blocks of evolution. As we will soon see, they are wrong on all counts. The mutations in the box cannot be dismissed, the mutations to the left of the box cannot necessarily all be eliminated by selection, and there is neither time nor selection power left for selecting the extremely rare beneficial mutations which might occur to the right of the near-neutral box.

Given the pivotal role beneficial mutations play in all evolutionary scenarios, I was puzzled as to why Kimura did not represent them in any way in his graph. In fairness I thought I should sketch them in. To the extent they occur, the distribution curve for beneficial mutations should be a reverse image of the deleterious mutations. Just like deleterious mutations, the overwhelming majority of the beneficial mutations should be nearly-neutral, being crowded toward the neutral mark. Crow (1997) clearly states this obvious fact, that the overwhelming majority of beneficial mutations should have very slight effects (for more detail on this see Appendix 5). However, the beneficial mutations are much rarer than deleterious mutations, and so their range and the area under their curve would be proportionally smaller. I have seen estimates of the ratio of deleterious-to-beneficial mutations which range from one thousand to one, up to one million to one. The best estimates seem to be one million to one (Gerrish and Lenski, 1998). The actual rate of beneficial mutations is so extremely low as to thwart any actual measurement (Bataillon, 2000, Elena et al, 1998). Therefore, I cannot draw a small enough curve to the right of zero to accurately represent how rare such beneficial mutations really are. Instead, I have just placed an easily visible triangle there (Figure 3d). Figure 3d is the most honest and true representation of the natural distribution of mutations (except that it is vastly too generous in terms of beneficial mutations). What is most interesting about this figure (it came as a shock to me) is to realize that essentially the entire range of all hypothetical beneficial mutations falls within Kimura's "effectively neutral" zone. That means that essentially all beneficial mutations (to the extent they actually happen), must be "un-selectable". So selection could never favor any such beneficial mutations, and they would essentially all drift out of the population. No wonder that Kimura preferred <u>not</u> to represent the distribution of the favorable mutations!

Figure 3d vividly illustrates why mutations cannot result in a net gain of information. As we will see more clearly in later chapters, selection cannot touch any of the mutations in the near-neutral box. Therefore, the very strong predominance of deleterious mutations in this box absolutely guarantees net loss of information. Furthermore, when mutation rate is high and reproductive rate is moderate or low, selection cannot even eliminate all the deleterious mutants to the left of the box. Last, we will see that constraints on selection even limit our ability to select for that extremely rare beneficial mutant that might lie to the right of the near-neutral box. **Everything about the true distribution of mutations argues against their possible role in forward evolution.**

Because beneficial mutations are so central to the viability of the Primary Axiom, I still need to say a little more about them. During the last century, there was a great deal of effort invested in trying to use mutation to generate useful variation. This was especially true in my own area – plant breeding. When it was discovered that certain forms of radiation and certain chemicals were powerful mutagenic agents, millions and million of plants were mutagenized and screened for possible improvements. Assuming the Primary Axiom, it would seem obvious that this would result in rapid "evolution" of our crops. For several decades this was the main thrust of crop improvement research. Vast numbers of mutants were produced and screened, collectively representing many billions of mutation events. A huge number of small, sterile, sick, deformed, aberrant plants were produced. However, from all this effort, almost no meaningful crop improvement resulted. The effort was for the most part an enormous failure, and was almost entirely abandoned. Why did this huge mutation/selection experiment fail – even with a host of Ph.D. scientists trying to help it along? It was because even with all those billions of mutations, there were no significant new beneficial mutations arising. The notable exceptions prove the point. For example, low phytate corn is the most notable example of successful mutation breeding. Such low phytate corn has certain advantages in terms of animal feed. The low phytate corn was created by mutatagenizing corn, and then selecting for strains wherein the genetic machinery which directs phytic acid production had been damaged. Although the resulting mutant may be desired for a specific agricultural purpose, it was accomplished through net loss of information (like the broken car alarm), and the loss of a biological function. Most of the other examples of successful mutation breeding are found within the area of ornamental plants - where dysfunctional anomalies are found to be novel and interesting to the eye. Examples of "useful" mutations within ornamental plants include sterility, dwarfing, mottled or variegated foliage, altered color patterns, or misshaped floral organs.

If essentially no truly beneficial mutations (i.e. resulting in more net information) could be recovered from this vast science-guided process, why do we think the identical process, in the absence of any guiding intelligence, would be more fruitful in nature? However, the very same scientists who failed at mutation/selection were *extremely* successful in crop improvement, when they abandoned mutation breeding and instead used the pre-existing natural variation within each plant species or genus. This would make sense if such pre-existing variation did not principally arise via mutation – but originally arose by *design*.

Bergman (2004) has studied the topic of beneficial mutations. Among other things, he did a simple literature search via Biological Abstracts and Medline. He found 453,732 "mutation" hits, but among these only 186 mentioned the word "beneficial" (about 4 in 10,000). When those 186 references were reviewed, almost all the presumed "beneficial mutations" were only beneficial in a very narrow sense - but consistently involved *loss-of-function* changes (hence loss of information). He was unable to find a single example of a mutation which unambiguously created new information. While it is almost universally accepted that beneficial (information-creating) mutations *must* occur, this belief seems to be based primarily upon uncritical acceptance of the Primary Axiom, rather than upon actual evidence. I do not doubt there *are* beneficial mutations, but it is clear they are exceedingly rare – much too rare for genome-building.

In conclusion, mutations appear to be overwhelmingly deleterious, and even when one may be classified as beneficial in some specific sense, it is still usually part of an over-all breakdown and erosion of information. As we will soon examine in greater detail, mutations, even when coupled with selection, cannot generally create new information. The types of variation created by mutation are more like the dings and scratches of life, and cannot be seen as life's spare parts (spare parts are designed). Mutations are the basis for the aging of individuals, and right now they are leading to our death - both yours and mine. Unless selection can somehow stop the erosion of information in the human genome, mutations will not only lead to our personal death - they will lead to the death of our species. We will soon see that natural selection must be able to simultaneously select against extremely large numbers of nearly-neutral nucleotide mutations, in order to prevent genomic degeneration.

atcgtacgtagcggctatgcgatgcaatgcatgctgctatatcgcatcgatatcggagatct
caccgtacgatttccgagagttaccaatcgatatggctatatccgcctttaggcgcctacac
atatttcatcgtacgcggctatgcgatgcaatgcgaatgctatatcgcatcgatatcgggac
gggacgatccacacttcggagagttaatacgatatggctataccggcctttaaagcctaca
atatattctcgtacgtagcaaaggctatgcgatgcaatgcgatgctctatatcgcatcgtaat
tcgggaatttgccgataatacgatatggctataccgccttaagcgttaactatcattcaacttt
atcgacgtagcgaagctatgcgatcatagcgatgctattcgatcgatactatcgggagcta
cgtacgctgatcggagagttaatacgatatggctatctccgcctttaagcgggctaacatat
attgtacgtagcggccccctaatgcgatgcaatcgcgatgctgatatcgacatcgatacga
atcgtacgtagcggctatgcgatgcaatgcatgctgctatatcgcatcgatatcggagatct
caccgtacgatttccgagagttaccaatcgatatggctatatccgcctttaggcgcctacac
atatttcatcgtacgcggctatgcgatgcaatgcgaatgctatatcgcatcgatatcgggatt
gggacgatccacacttcggagagttaatacgatatggctataccggcctttaaagcctaca
atatattctcgtacgtagcaaaggctatgcgatgcaatgcgatgctctatatcgcatcgtaat
tcgggaatttgccgataatacgatatggctataccgccttaagcgttaactatcattcaacttt
atcgacgtagcgaagctatgcgatcatagcgatgctattcgatcgatactatcgggagcta

Figure 2.

The genome appears to us as a linear array of letters: A, T, C, G. The actual genome is 3 million fold greater than the sequence shown above. To view just half of your own genome, you would have to view 10 nucleotides every second for 40 hours per week, for 40 years! The apparent simplicity of this language system is deceptive. A higher genome, almost certainly, must comprise a great deal of data compression (see Chapter 9), as well as a great deal of non-linear information. Except for certain short portions, we cannot view the genome as simply a linear text, like a book. Much of the information content is probably found in 3-dimensional structures, as is the case with folded proteins.

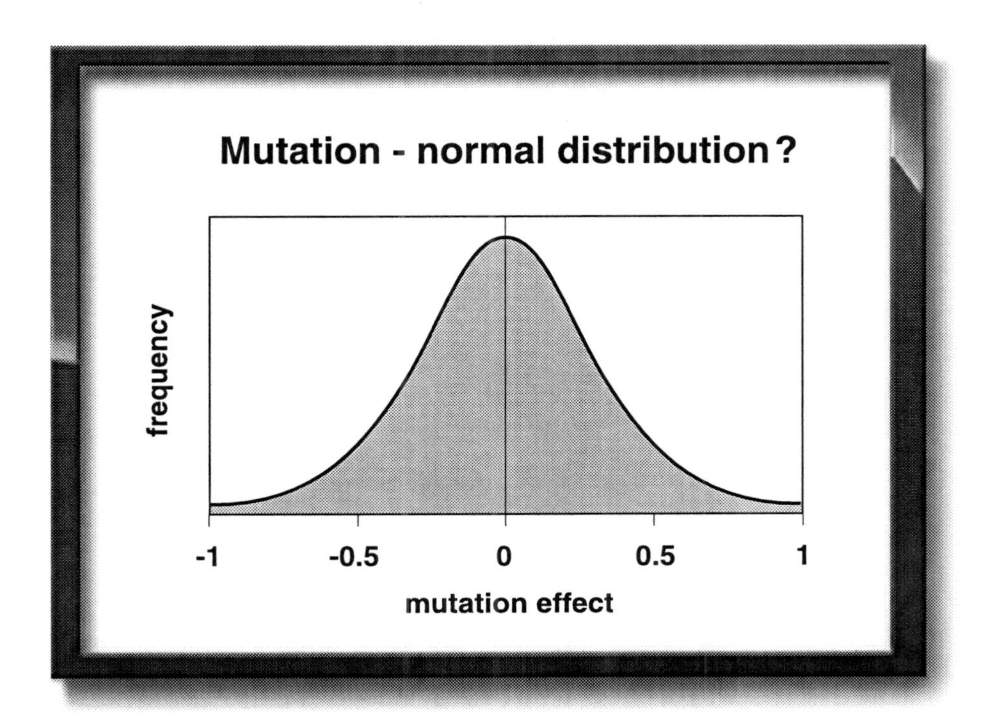

Figure 3a.

The naïve view of mutations would be a bell-shaped distribution, with half of all mutations showing deleterious affects on fitness (left of center), and half showing positive effects on fitness (right of center). With such a distribution, it would be easy to imagine selection removing bad mutations and fixing good mutations - resulting inevitably in evolutionary progress. However, we know this is a false picture.

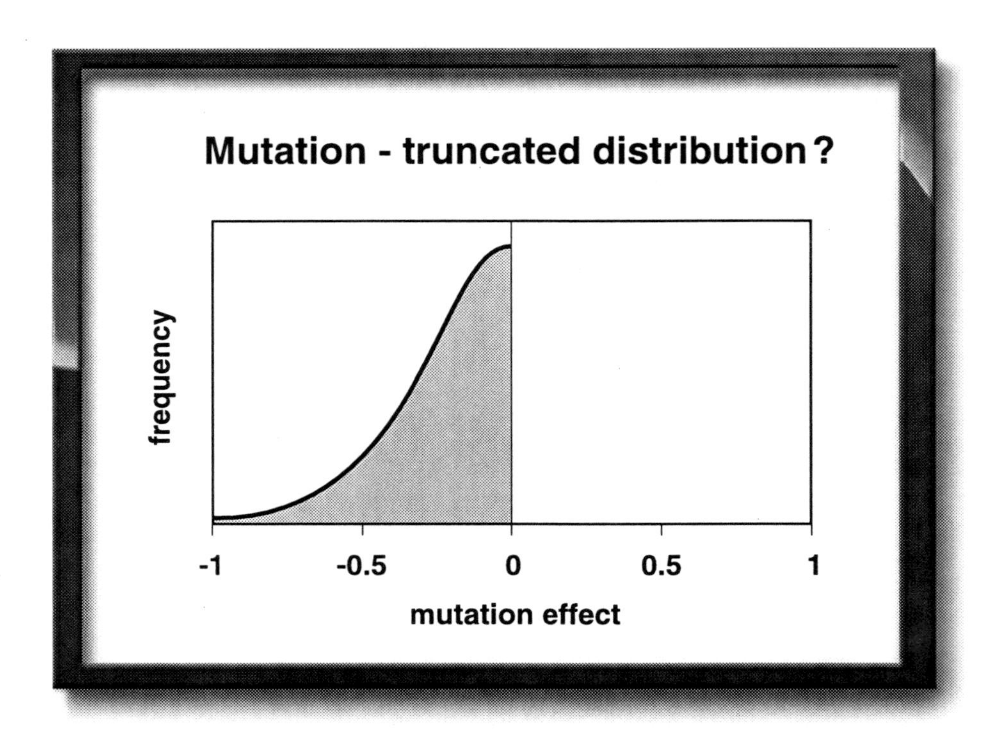

Figure 3b.

Population geneticists know that essentially all mutations are deleterious, and that mutations having positive effects on fitness are so rare as to be excluded from such distribution diagrams. This creates major problems for evolutionary theory. But this picture is still too optimistic.

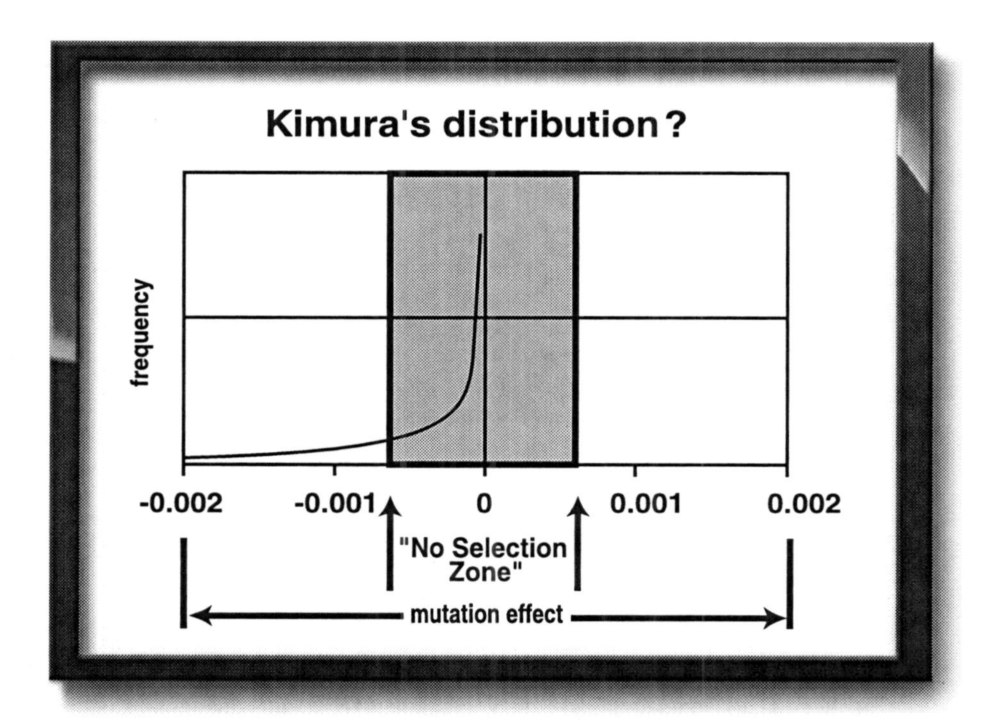

Figure 3c.

Population geneticists know that mutations are strongly skewed toward neutral. Just like in an instruction manual, a few misspellings will be lethal - but most will be *nearly harmless*. The nearly-neutral mutations create the biggest problems for evolutionary theory. This diagram is adapted from a figure by Kimura (1979). Kimura is famous for showing that most mutations are nearly-neutral, and therefore are not subject to selection. Kimura's "no-selection zone" is shown by the box.

The general shape of this curve is important, but the precise mathematical nature of this curve is not. While Ohta feels the mutation distribution is exponential, Kimura feels it is a 'gamma' distribution (Kimura, 1979). However, regardless of which specific mathematical formulation best describes the natural distribution of mutation effects, they all approximate the picture shown above.

*For your possible interest, geneticists agree that the frequency of highly deleterious mutations is almost zero (not shown - off the chart), while 'minor' mutations are intermediate in frequency (i.e. - the left portion of chart, and off chart). Minor mutations are believed to outnumber major mutations by about **10-50** fold (Crow, **1997**), but near-neutrals vastly outnumber them both.*

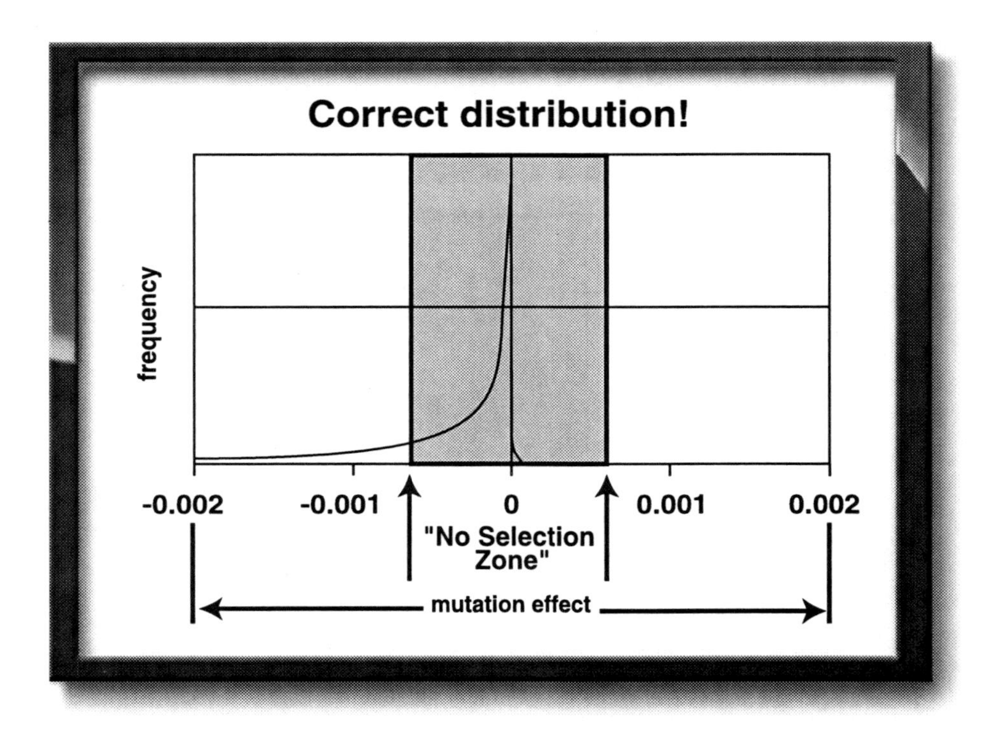

Figure 3d.

Kimura's Figure (3c) is still not complete. To complete the figure we really must show where the beneficial mutations would occur (as they are critical to evolutionary theory). Their distribution would be a reverse image of Kimura's curve, but reduced in range and scale, by a factor of somewhere between ten thousand to one-million. Because of the scale of this diagram, I cannot draw this part of the mutation distribution small enough, so a relatively large triangle is shown instead. Even with beneficial mutations greatly exaggerated, it becomes obvious that essentially all beneficial mutations will fall within Kimura's "no-selection zone". This completed picture, which is correct, makes progressive evolution, on the genomic level, virtually impossible.

How much mutation is too much?

Newsflash - Human mutation rates are much too high.

For many decades geneticists have been worried about the impact of mutation on the human population (Muller 1950, Crow, 1997). When these concerns first arose, they were based upon an estimated rate of deleterious mutation of *0.12 to 0.30 mutations per person per generation* (Morton, Crow and Muller, 1956). Since that time there have persisted serious concerns about accumulating mutations in man leading to a high "genetic load" - and a generally degenerating population. There has also been a long-standing belief that if the rate of deleterious mutations approached *one deleterious mutation per person per generation*, long-term genetic deterioration would be a certainty (Muller, 1950). This would be logical, since selection must eliminate mutations as fast as they are occurring. We need to prevent mutant individuals from reproducing, but we also need to leave enough remaining people to procreate and produce the next generation. By this thinking, deleterious mutations in man must actually be kept below one mutation for every three children - if selection is to eliminate all the mutations and still allow the population to reproduce. This is because global fertility rates are now less than 3 children for every 2 adults - so only one child in three could theoretically be selectively eliminated. For these reasons, geneticists have been naturally very eager to discover what the human mutation rate really is!

One of the most astounding recent findings in the world of genetics is that the human mutation rate (just within our reproductive cells) is at least 100 nucleotide substitutions (misspellings) per person per generation (Kondrashov, 2002). Other geneticists would place this number at 175 (Nachman and Crowell, 2000). These high numbers are now widely accepted within the genetics community. Furthermore, Dr. Kondrashov, the author of the most definitive publication, has indicated to me that 100 was only his *lower estimate* – he believes the actual rate of point mutations (misspellings) per person may be as high as 300 (personal communication). Even the lower estimate, 100, is an amazing number, with profound implications. When an earlier study revealed that the human mutation rate might be as high as *30*, the highly distinguished author of that study, concluded that such a number would have *profound* implications for evolutionary theory (Neel et al., 1986). But the actual number is now known to be 100-300! Even if we were to accept the lowest estimate (100 mutations), and further assumed that 97 % of the genome is perfectly neutral junk, this would still mean that at least 3 additional deleterious mutations are occurring per person per generation. So *every one* of us is a mutant, many times over! What type of selection scheme could possibly stop this type of loss of information? As we will see - given these numbers, there is no realistic method to halt genomic degeneration. Since the portion of the genome that is recognized as being truly functional is rapidly increasing, the number of mutations recognized as being actually deleterious is also rapidly increasing. If all the genome proves functional, then every one of these 100 mutations per person is actually deleterious. Yet even this number is too small, firstly because it is only the lowest estimate, and secondly because it only considers point mutations (misspellings). Not included within this number are the many other types of common mutations - such as deletions, insertions, duplications, translocations, inversions, and all mitochondrial mutations.

To appreciate to what extent we are still underestimating the mutation problem, we should first consider the types of mutation which fall outside the scope of the normal 'point mutation' counts. Then we need to consider what portion of the whole genome is really functional, and not "junk".

Within each cell are sub-structures called mitochondria, which have their own small internal genome (about 16,500 nucleotides), which is inherited only through the mother. However, because the mitochondrial genome is highly polyploid (hundreds of copies per cell), and because the mitochondrial mutation rate is extremely high, there are still a large number of mitochondrial mutations that must be eliminated each generation - to halt degeneration. The human mitochondrial mutation rate has been estimated to be about 2.5 mutations, per nucleotide site, per million years (Parsons et al, 1997). Assuming a generation time of 25 years and a mitochondrial genome size of 16,500 - this approaches *one mitochondrial mutation per person per generation within the reproductive cell line.* Mitochondrial mutations, just by themselves, probably put us over the theoretical limit of one mutation per three children! Even if the mutation rate is only 0.1 per person, we would have to select away a very substantial portion (10%) of the human population, every generation, just trying to halt mitochondrial genetic degeneration. Yet this would still leave the 100-300 nuclear mutations per person per generation (as discussed above) accumulating - unabated. High rates of mitochondrial mutation are especially problematic in terms of selection (Chapters 4 and 5), because of lack of recombination ("Muller's ratchet" - Muller, 1964), and lower effective population size (only women pass on this DNA, so selection can only be applied to half the population).

The most rapidly mutating regions of the human genome are within the very dynamic micro-satellite DNA regions. These unique regions mutate at rates nearly 1 million-fold above normal, and are not included in normal estimates of mutation rate. Yet these sequences are found to have biological impact, and their mutation results in many serious genetic diseases (Sutherland and Richards, 1995). It is estimated that for every "regular" point mutation, there is probably at least one micro-satellite mutation (Ellegren, 2000). This effectively doubles the mutation count per person per generation, from 100-300 to 200-600.

In addition to nuclear point mutations, mitochondrial mutations, and micro-satellite mutations, there are a wide variety of more severe chromosomal mutations - called macro-mutations. These include deletions and insertions. According to Kondrashov (2002), such mutations, when combined, add another 4 macro-mutations for every 100 point mutations (this estimate appears to consider only the smallest of macro-mutations, and excludes the insertions/ deletions affecting larger regions of DNA). Although there may be relatively few such mutations (*only* 4-12 per person per generation), these "major" mutations will unquestionably cause much more genomic damage, and so would demand higher priority if one were designing a selection scheme to stop genomic degeneration. Macro-mutations can affect any number of nucleotides – from one to a million - even as we might accidentally delete a letter, a word, or even an entire chapter from this book. These relatively few macro-mutations are believed to cause 3- to 10-fold more sequence divergence than all the point mutations combined (Britten, 2002; Anzai, 2003). This brings our actual mutation count per person per generation up to about 204 - 636. But if we factor in the fact that macro-mutations can change 3- to 10-fold more nucleotides than all point mutations combined, our final tally of nucleotide changes per person could come up to as high as 612-6,360 per person per generation! These numbers are mind-boggling! Yet even these numbers may *still* be too low - we have not yet considered inversions and translocations. Furthermore, evolutionary theorists are now invoking extremely high inter-genic *conversion* rates, which could double these numbers again. Wow! Do you recall the beginning of this chapter, where we learned that the famous geneticist Muller considered that a human mutation rate of 0.5 per person or higher, would doom man to rapid genetic degeneration? Although we do not know the precise human mutation rate, there is good reason to believe that there are more than 1,000 nucleotide changes in every person, every generation (see Table 1). To be exceedingly generous, for the rest of this book I will use the most conservative number being referred to in the literature today - *'just'* 100 mutations per person per generation (except where otherwise specified). However, please note that this is only a fraction of the true number, and this number excludes the most destructive classes of mutations.

Of all these mutations - what percent are truly neutral? In the last few years there has been a dramatic shift in the perceived functionality of most components of the genome. The concept of "junk DNA" is quickly disappearing. In fact, it is the "junk DNA" (non-protein-coding DNA), which appears to be *key* to encoding biological complexity (Taft and Mattick, 2003). The recent Taft and Mattick study strongly suggests that the more "junk" - the more advanced is the organism. So mutations within "junk DNA" can hardly be assumed to be neutral!

Approximately 50% of the human genome is now known to be transcribed into RNA (Johnson et al., 2005). At least half of all this transcribing DNA appears to be *transcribed in both directions* (Yelin et al., 2003)! So all of this DNA is not only functional – but much of it may be *doubly* functional. While only a small fraction of the genome directly encodes for proteins, every protein-encoding sequence is embedded within other functional sequences that *regulate* the expression of such proteins. This includes promoters, enhancers, introns, leader sequences, trailing sequences, and sequences affecting regional folding and DNA architecture. I do not believe any serious biologist now considers introns (which comprise most of a typical genic region) as truly neutral "junk". In fact, many of the most strongly *conserved* (essential and invariant) sequences known, are found within introns (Bejerano et al., 2004). While a typical protein-coding sequence may only be 3,000 nucleotides long or less, the typical "whole gene" that controls the expression of that protein can be in the range of 50,000 nucleotides long. Since there are 20,000 - 40,000 protein-encoding genes (estimates greatly vary), if we include all their associated nucleotides (50,000 per gene), the true complete genes could easily account for over 1.5 billion nucleotides. This is fully half the genome. In addition, a whole new class of genes has been discovered which do not encode proteins, but encode functional RNAs. Such genes have escaped recognition in computer searches for protein-coding sequences, and so have been overlooked as true genes. But they are true genes, and they probably comprise a large part of the genome (Mattick, 2001; Dennis, 2002; Storz, 2002). They are just now being discovered - within DNA regions that were previously dismissed as "junk". In addition, two independent studies have shown extensive sequence functionality within the large regions *between* genes (Koop and Hood, 1994; Shabalina et al., 2001). Previously, such regions had also been assumed to be junk. Pseudogenes, long considered dead duplicated genes, have recently been shown to be functional (Hirotsune et al., 2003; Lee, 2003). Pseudogenes seem to be designed to make regulatory RNA molecules (see Chen et al. 2004), rather than proteins, so they are not "dead fossils". As I will discuss in more detail elsewhere, there even appear to be diverse cellular functions for the much-maligned "selfish genes", sometimes called "parasitic DNA sequences", also called "transposable elements". These elements appear to have multiple, and extremely important functions within the cell, including the control of chromosome pairing (Hakimi et al. 2002), and DNA repair (Morrish, et al., 2002). Repetitive DNA , including satellite DNA, long considered junk, has been shown to be essential to genome function, and comprise such essential genomic structures as centromeres and telomeres (Shapiro and Sternberg, 2005). Lastly, there are fundamental genome-wide structural patterns, which virtually permeate every portion of the genome - such as isochores (GC rich areas -Vinogradov, 2003), genome-wide 'word' patterns (Karlin, 1998) and nucleosome binding sites (Tachida, 1990). These genome-wide patterns appear crucial to cell function, and suggest functionality throughout the entire genome. For example, nucleosome binding (crucial to chromosome structure and gene regulation) appears to be specified by di-nucleotide patterns that repeat every 10 nucleotides (Sandman et al., 2000). It appears that one-fifth of the genome may be functional and essential - *just for the purpose of specifying nucleosome binding sites* (Tachida, 1990). **It is becoming increasingly clear that most, or all, of the genome is functional. Therefore, most, or all, mutations in the genome must be deleterious.**

On a per person basis, 100 mutations represent a loss of only a miniscule fraction of the total information in our genome (the genome is huge). However, the real impact of such a high mutation rate will be at the population level, and is primarily expressed with the passage of time. Since there are six billion people in the world, and each person has added an average of 100 new mutations to the global population, our generation alone has added roughly 600 billion new mutations to the human race. If we remember that there are only three billion nucleotide positions in the human genome, we see that in our lifetime there have been about 200 mutations for every nucleotide position within the genome. Therefore, every possible point mutation that *could* happen to the human genome *has* happened many times over - just during our lifetime! Because of our present large population size, humanity is now being flooded by mutations like never before in history. The consequences of most of these mutations are not felt immediately, but will become manifested in coming generations.

As we will be seeing, there is no selection scheme that can reverse the damage that has been done during our own generation - even if further mutations could be stopped. No amount of selection can prevent a significant number of these mutations from drifting deeper into the population and consequently causing permanent genetic damage to the population. Yet our children's generation will add even more new mutations - followed by the next and the next. This degenerative process will continue into the foreseeable future. We are on a downward slide that cannot be stopped.

When selection is unable to counter the loss of information due to mutations, a situation arises called "error catastrophe". If not rapidly corrected, this situation leads to the eventual death of the species - extinction. In its final stages, genomic degeneration leads to declining fertility, which curtails further selection (selection always requires a surplus population - some of which can then be eliminated each generation). Inbreeding and genetic drift must then take over entirely - rapidly finishing off the genome. When this point is reached, the process becomes an irreversible downward spiral. This advanced stage of genomic degeneration is called "mutational meltdown" (Bernardes, 1996). Mutational meltdown is recognized as an immediate threat to all of today's endangered species. The same process appears to potentially be a theoretical threat for mankind. What can stop this process?

Mutation Type	Mutations per Person	Nucleotides changed/person
1. mitochondrial mutations[a]	<1	<1
2. nucleotide substitutions[b]	100-300	100-300
3. satellite mutations[c]	100-300	100-300
4. deletions[d]	2-6 (plus)	300-3000
5. duplications / insertions[e]	2-6 (plus)	300-3000
6. inversions / translocations[f]	numerous	thousands?
7. conversions[g]	thousands?	thousands?
total/person/generation[h]	**>1,000?**	**thousands!**

Table 1.

There are many types of mutation, which act as sources of heritable genetic change. Unfortunately, every single class of mutation results in a net loss of information. Mitochondrial mutation is the least significant source of human mutation – it produces less than one new mutation per person. Yet even a fraction of one mitochondrial mutation per person has prompted one evolutionist to comment: "***We should increase our attention to the broader question of how (or whether) organisms can tolerate, in the sense of evolution, a genetic system with such a high mutational burden.***" **(Howell et al., 1996).** Now, consider all the types of mutation combined!

[a] *Mitochondrial mutation rate estimates vary, but can approach **0.5** per person (Parsons et al., **1997**).*

[b] *Nuclear substitutions are hard to measure, but Kondroshov (**2002**) has estimated **100** per person. In personal communication he has indicated this may actually be **300**.*

[c] *Normal estimates of nucleotide substitutions would not include mutational hotspots such as microsatellites. Microsatellite mutation rates have been estimated to be roughly equal to all point mutations rates.*

[d,e] *Kondrashov (**2002**) estimated that deletions plus insertions occur at a combined rates of about **4-12%** of the point mutations – or about **2-6%** each. However, he seemed to limit his estimate to only small inserts and deletions, so the actual number may be higher. Because mutations and insertions can be very large, their total effect is believed to be **3-10** fold greater than all point mutations, in terms of total nucleotides changed.*

[f] *The actual rate of chromosomal rearrangements is unknown. Evolutionary assumptions about the recent divergence of chimp and man require high rates of such changes. These changes can affect very large pieces of DNA, and so for the evolutionary scenario to work, many thousands of nucleotides on average, must move in this way every generation.*

[g] *The actual rate of inter-genic conversion is unknown, but evolutionary assumptions require extremely high rates of gene conversion between different loci – many thousands per person per generation.*

[h] *The total number of mutations can only be estimated in a very crude way, but it should be very clear that the number of all types of new mutations, including conversions, must be over **1,000** per person. These mutations, which include many macro-mutations, must clearly change many thousands of nucleotides per person per generation.*

All-powerful Selection to the rescue?

Newsflash - Selection capabilities are very limited.

The consensus among human geneticists is that at present the human race is degenerating genetically, due to rapid mutation accumulation and relaxed natural selection pressure (Crow, 1997). These geneticists realize that there is presently a net accumulation of mutations in the population, which is occurring at a much higher rate than was previously thought possible. Geneticists widely agree that essentially all of these mutations are either neutral or deleterious (if any are beneficial, they are considered so rare as to be entirely excluded from consideration). Subsequently, they realize that genetic information is currently being lost, which must eventually result in reduced fitness for our species. This decline in fitness is believed to be occurring at 1-2% per generation (Crow 1997) (see Figure 4). All this is happening on the genetic level, even though on the social level medical and technical advances are actually still increasing our average life spans. Hence human geneticists would probably all agree that eventually selection must be increased if we are to stop genetic degeneration. However, there are essentially no public statements to this effect --- imagine the profound political ramifications of such statements!

This acknowledged problem raises an interesting question - "How much selection would be required to completely halt genetic degeneration?" Or perhaps the question should really be - "Can such degeneration be halted at all?"

For many people, including many biologists, natural selection is something like a magic wand. Simply by invoking the words "natural selection" – there seems to be no limit to what one can imagine accomplishing. This extremely naïve perspective toward natural selection is pervasive. Even as a plant geneticist and plant breeder, I still had an unrealistic conception of how selection was really operating in nature, and I had a very naïve idea about how selection might work at the highest level – on the level of the whole genome. For the most part, the only scientists who have actually seriously analyzed what selection can do (and cannot do) on the genomic level, are a small number of population geneticists – an exceedingly specialized group. Population genetics is a field that is extremely theoretical and mathematical. Theoretical mathematicians are constrained (completely) by their *axioms* (assumptions), upon which they choose to build their work. The entire field of population genetics was developed by a small, tightly knit group of people who were utterly and radically committed to the Primary Axiom. Today, it is still a very small field, still exclusively populated by "true believers" in the Primary Axiom. These people are extremely intelligent, but are totally and unconditionally bound to the Primary Axiom. For the most part, other biologists do not even understand their work - but accept their conclusions "by faith". Yet it is *these* same population geneticists themselves who have exposed some of the most profound limitations of natural selection (see Appendix 1). Because natural selection is *not* a magic wand but is a very real phenomenon, it has very real capabilities and very real *limitations*. It is not "all-powerful".

The Most Basic Problem -
The Princess and the Nucleotide Paradox

Natural selection has a very fundamental problem. This problem involves the enormous chasm that exists between **genotypic change** (a molecular mutation) and **phenotypic selection** (a whole organism's reproduction). There needs to be selection for billions of almost infinitely subtle and complex genetic differences on the molecular level. But this can only be done by controlling reproduction on the level of the whole organism. When Mother Nature selects for or against an individual within a population, she has to accept or reject a complete set of 6 billion nucleotides - all at once. It's either take the whole book - or have nothing of it. In fact, Mother Nature really **never even sees the individual nucleotides** – she just sees the whole organism. So she never has the luxury of seeing, or selecting for, any particular nucleotide. We start to see what a great leap of faith is required to believe that by selecting or rejecting a whole organism, Mother Nature can precisely control the fate of billions of individual misspellings within the assembly manual.

The problem of *genotypic change* versus *phenotypic selection* is very much like the problem of the *Princess and the Pea.* The children's story of the Princess and the Pea involves the discovery of a Princess' true and royal character. She is discovered to be a Princess by the fact that she cannot sleep - because even through 13 mattresses, she feels a pea beneath her bed. Children are entertained by this story because it is so silly. Royalty or not – no one can feel a pea through 13 mattresses! But our genetic problem is actually a much more difficult situation. Our Princess (natural selection) essentially needs to read extensive books written in Braille through the 13 mattresses, so she can precisely identify which books have the least errors in them! It makes a great fairy tale – but who would believe it as the underlying process which explains life? This whole problem can be called the *Princess and the Nucleotide Paradox*, which is whimsically illustrated in Figure 5.

To be fair, there are a few mutations that have a much bigger effect than a single Braille letter in our example. A few rare mutations have profound biological effects – acting more like a bowling ball under the mattress. Natural selection against these types of major mutations is an obvious "no-brainer". But the "bowling ball" (semi-lethal) mutations are extremely rare, and such nucleotides carry only a miniscule amount of the total information in the genome. Most of the information in the genome is carried by nucleotides whose effects are actually much more subtle than even the Braille letters in our whimsical example. It is the origin and maintenance of all *those* nucleotides that we are trying to understand.

The gap between molecular mutations and a whole organism's reproduction is very profound. Part of this gap involves size (if we were to make a nucleotide as big as a pea, proportionately we would have to make a real-life princess roughly 10,000 miles tall). Moreover, standing between a nucleotide and an individual organism are many different *levels of organization*. For example, a single nucleotide may affect a specific gene's transcription, which may then affect mRNA processing, which may then effect the abundance of a given enzyme, which may then affect a given metabolic pathway, which may then affect the division of a cell, which may then affect a certain tissue, which may then affect a whole organism, which may then affect the probability of reproduction, which may then affect the chance that that specific mutation gets passed on to the next generation. Massive amounts of uncertainty and dilution are added at each organizational level, resulting in massive increase in "noise", and loss of resolution. There must be a *vanishingly small* correlation between any given nucleotide (a single molecule), and a whole organism's probability of reproductive success! The nucleotide and the organism are very literally worlds apart. Our Princess (i.e. natural selection on the level of the whole organism), in fact, has to perceive differences which are just above the *atomic level.*

We do not generally see individual pixels on our television - so imagine the difficulty of trying to select a specific TV set at the store by trying to evaluate the quality of each separate pixel (by eye), on all the various TV sets available. But it's really much worse than this. In a biological system, we are talking about pixels, within pixels, within pixels, within pixels. We are talking about a very long chain of events separating the direct effect of a given nucleotide and very remote consequences on the whole organism level. There is a logarithmic dilution at each step – at each level there is an order of magnitude loss of resolution and correspondence. It is like measuring the impact of a butterfly's stroke - on a hurricane system which is a thousand miles away. It is a little like trying to select for a specific soldier, based upon the performance of his army. This whole picture is totally upside down! Yet this is the essence of the Primary Axiom! The Primary Axiom sees a human genome (6 billion nucleotides), and imagines that each unit is selected for (or not) individually, based merely upon a limited amount of reproductive sieving on the level of the whole organism. As we will be seeing, this is totally impossible for many reasons.

To better understand the nature of The Princess and the Nucleotide Paradox, let's imagine a new method for improving textbooks. Let's start with a high school biochemistry textbook (equivalent to let us say, a very simple bacterial genome), and let's begin introducing random misspellings, duplications, and deletions. Each student, across the whole country, will get a slightly different textbook – each containing its own set of random errors - approximately 100 new errors per text. At the end of the year, we will test all the students, and we will only save the textbooks from the students with the best 100 scores. Those texts will be used for the next round of copying, which will introduce new errors...etc. Can we expect to see a steady improvement of textbooks? Why not? Will we expect to see a steady improvement of average student grades? Why not?

Most of us can see that in the above example, essentially **none** of the misspellings in the textbook will be beneficial. More important to my point, there will be no meaningful correlation between the subtle differences in textbooks and a student's grades. Why not? Because every textbook is approximately equally flawed, and the differences between texts are too subtle to be significant - in light of everything else. What do I mean by "everything else"? I mean that a student's grade will be determined by so many other important variables – the student's own personal abilities, all the different personal situations, teachers, classrooms, other kids, motivation, home life, romantic life, lack of sleep, "bad luck", etc. All these other factors (which I will call *noise*) will override the effect of a few misspellings in the textbook. If the student gets a high grade on the test, it is not because his text had slightly fewer errors - but primarily for all those other diverse reasons.

What will happen if this mutation/selection cycle continues unabated? The texts *will* obviously degenerate over time, and average student scores *will* eventually also go down. Yet this absurd mutation/selection system is a very reasonable approximation of the Primary Axiom of biology. It very obviously will fail to improve or even maintain grades - for many reasons. The most fundamental reason why this type of selection fails is the incredibly weak relationship between individual letters in the text and the over-all performance of the student. The correlation will be essentially zero. This is an excellent illustration of the Princess and the Nucleotide Paradox. If this scenario seems absurd to you, try to understand one more thing. The Primary Axiom claims that this very same mutation/selection system is actually what *wrote the entire biochemistry textbook in the first place.* There was never any intelligent agent acting as author, or even as editor.

The problem of the Princess and the Nucleotide Paradox becomes even greater when we understand the phenomenon of *homeostasis.* Homeostasis is the natural phenomenon wherein all living things *self-regulate* themselves to stay the same, even as circumstances change. A good example would be warm-blooded animals in a cold climate. Homeostasis results from an incredibly complex network of sensors and regulators within each cell. Although it is too complex to explain (or even analyze in detail), it is universally agreed that it is both operational and highly effective in all life systems. The phenomenon of homeostasis is a little like having a super-duper, self-adjusting mattress. If a tennis ball is put beneath this mattress, the mattress will automatically adjust itself via a myriad of complex mechanical mechanisms to effectively level the sleeping surface. But in real biology, it is worse – it is more like 13 self-adjusting mattresses on top of each other (homeostasis operates at every level of biological organization - from molecules to man). This makes things much more difficult for our Princess – who needs to sense the pea, or more accurately, read the Braille, through the mattresses.

When Mendel's genetic principles were "re-discovered" almost 50 years after Darwin, geneticists started to realize that there must be large numbers of hereditary units segregating within any given population. It was after this time that "population genetics" was born. If the number of hereditary units was very large – they realized they had a problem. Although they did not speak of it as such, it was essentially what I am now calling the Princess and the Nucleotide Paradox. The early population geneticists, who were all philosophically committed Darwinists, realized they had to devise a way to overcome the Princess and the Nucleotide Paradox in order to make Darwinian theory appear genetically feasible*. To overcome the problem, they very cleverly transferred the *unit of selection* from the whole organism to the genetic unit (i.e. the gene or nucleotide). To do this they had to re-define a population of people as being essentially a "pool of genes". In this way, they could claim that real selection was operating at the level of each individual nucleotide within the pool - not really on people at all. Each individual nucleotide could be envisioned as being independently selected for, or against (or neither). This made it very easy to envision almost any evolutionary selection scenario, no matter how complex the biological situation. This effectively removed the mattresses from under the Princess – as if she could suddenly feel each pea, and could even read each Braille letter – directly! This was an incredibly effective way to obscure the entire problem. Indeed, Darwinism would have died very naturally at

* *"Haldane.. intended..., as had Fisher... and Wright... to dispel the belief that Mendelism had killed Darwinism... Fisher, Haldane, and Wright then quantitatively synthesized Mendelian heredity and natural selection into the science of population genetics." The Origin of Theoretical Population Genetics. Provine, 1971.*

this point in time, except for this major intellectual invention (Provine, 1971).

There is one serious problem with re-definition of the problem in this way. That problem is that the new picture is categorically false. Populations are not even remotely like pools of genes, and selection is ***never, ever*** for individual nucleotides. To justify this radical new picture of life, the theorists had to axiomatically assume a number of things - which were in fact all known to be clearly false. For example, they had to assume that all genetic units could assort independently – so each nucleotide would be inherited independently – as if there were no linkage blocks (totally false). Likewise, they had to assume no epistasis – as if there were no interactions between nucleotides (totally false). They also typically assumed essentially infinite population sizes (obviously false), they usually implicitly assumed unlimited time for selection (obviously false), and they generally assumed the ability to select for unlimited numbers of traits simultaneously (which we will show to be false). So from the very beginning of population genetic theory many unrealistic and unreasonable assumptions needed to be made, to make the model appear even feasible.

On this very false foundation was built the theoretical pillars of modern population genetics. The theorists' models did not match biological reality, but these men had an incredible aura of intellectual authority, their arguments were very abstract, and they used highly mathematical formulations which could effectively cow most biologists. Furthermore, most biologists were also committed Darwinists, and so were philosophically in agreement with the population geneticists. So most biologists were more than happy to go along for the ride, even if the story did not quite make sense. In fact, the early population geneticists quickly became the "darlings of science" and were essentially idolized. I remember my own graduate level classes in population biology – and my own naive and meek acquiescence in accepting the very unnatural re-definition of life as "pools of genes". I remember not quite getting it, and assuming the problem was just with me (although all the other students in my class seemed to have the same problem). Since I "knew evolution was true", it did not really matter if I was not *quite smart enough* to really grasp the idea of life as pools of nucleotides. If "the gurus of population genetics" were saying it was true – who was I to argue? Even if their premises really were false (such as independent assortment of nucleotides), their conclusions must still doubtless be true - they were geniuses! Even though I was actually one of the more free-thinking students, I still swallowed the Princess and the Nucleotide story with essentially no resistance.

What is the biological reality – apart from ideology? The reality is that selection acts on the level of the organism, not on the level of the nucleotide (see Crow and Kimura, 1970, p. 173). Human genes never exist in "pools", they only exist in massive clusters, within real people. Each nucleotide exists intimately associated with all the other nucleotides within a given person, and they are only selected or rejected as a set of 6 billion. The phenomenon of linkage is profound and extensive - as we will be learning. No nucleotide is <u>ever</u> inherited independently. Each nucleotide is intimately connected to its surrounding nucleotides – even as each letter on this page is specifically associated with a specific word, sentence, paragraph, and chapter. This book was not produced by a selective system like a giant slot machine - where each letter is selected independently via a random process. Each letter was put in place by design, as part of something greater – as part of a specific word, a sentence, a paragraph and a chapter. This is also true of nucleotides – they only exist and have meaning in the *context* of other nucleotides (which is what we call epistasis). We now know that human nucleotides exist in large linked clusters or blocks, ranging in size from 10,000 to a million. These linkage blocks are inherited as a single unit, and never break apart. This totally negates one of the most fundamental assumptions of the theorists – that each nucleotide can be viewed as an individually selectable unit. Since the population genetics model of life (as pools of genes) is categorically false, the Princess and the Nucleotide Paradox remains entirely unresolved. This should be an enormous embarrassment to the entire field of population genetics. On a practical level – it means natural selection can never create, or even maintain, specific nucleotide sequences.

Not only is the Princess and the Nucleotide Paradox unresolved, we now know that the problem is vastly worse than the early population geneticists could have imagined. We have now learned that the size and complexity of the genome is vast (representing the extent of the Braille books), that homeostasis is extensive (representing the thickness of the mattresses), and there are many more levels of organization separating the genotype and the phenotype (representing the number of mattresses). In addition to all these problems – we will have to wait until Chapter 6 to fully understand the problem of biological *noise* (it turns out the mattresses themselves are full of pea-sized lumps). So we should be able to see that the Princess and the Nucleotide Paradox really is a show-stopper. The Primary Axiom fails - at this first and most basic level. Any child should be able to see it – although many adults are "too well educated" to see it. Yet the paradox of the

Princess and the Nucleotide is just the beginning of the problems for the Primary Axiom. For the purpose of further discussion, and for the rest of this book, I will be happy to give the theorists their model of life as "pools of genes", and the idea of selection on the level of individual nucleotides. I can agree to do this, because as we will see, there are many other problems which still fully discredit the Primary Axiom. But just for the record, the Princess and the Nucleotide Paradox is in itself sufficient basis for rejecting the Primary Axiom (see Appendix 3).

Three Specific Selection Problems

To understand the basic issues of genomic selection (which will soon become fairly complex), let us look at this question using the simplest case - a single point mutation which has accumulated within the human population to the extent that 50% of all people bear this mutation. What type of selection is required to eliminate this type of mutation, and what are the critical factors? For simplicity, we will be assuming the mutation is dominant (almost all mutations are recessive, which makes selection much more difficult). At first glance, the problem seems very easily solved. We could eliminate all these mutants in a single generation, if we could: 1) afford to lose 50% of the breeding population; 2) if we could clearly identify every person carrying the mutation; and 3) if we could prevent 100% of the carriers from reproductive mating. So what are the problems?

1. Cost of selection. The problem of the cost of selection was first described by Haldane (1957), and later validated and expanded upon by Kimura and Ohta (1971), and Kimura (1983). It has been further clarified by ReMine (1993, 2005). All selection involves a biological cost – meaning that we must remove (or "spend") part of the breeding population. This is the essence of selection! In the current example, we should ask - "Can we really afford to spend 50% of humanity, preventing half the people from reproducing, so that we can make rapid selective progress?" Given humanity's fertility levels (globally – less than 3 children per 2 adults), if we eliminate 50% of our population for the purpose of selection the population size will be reduced by 25%. Obviously, each two adults need to have at least two reproducing children to maintain the population size. Furthermore, not all children will go on to reproduce (for many reasons - accidental death, personal choice, etc.). Therefore, considerably more than two children per two adults are needed to keep the population viable (given our low fertility, if three children per two adults were needed for population continuity, zero selection would be possible). For these reasons, substantially less than one child in three is available to be spent for selection purposes. Haldane (1957) believed that only 10% of a typical natural human population can realistically be spent for selection purposes. So if 50% of the population *was* removed for purposes of selection every generation, then the human population would shrink rapidly - eventually leading to our extinction. Therefore, in the above example, elimination of all the mutant individuals in one generation would not really be reasonable. However, doing this same amount of selection in two generations *might* be reasonable - since only 25% of the population would be spent for selection, per generation.

The purpose of this simple illustration is to show that while selection works, there are clear limits in terms of how intense our selection can be, and we must understand that every selective event has a biological <u>cost</u> in terms of non-reproduction. This becomes a major issue when selecting against many different mutations simultaneously. For the human population, it becomes clear that the maximum part of our population which can be "spent" for all selection purposes is considerably less than 33% - and according to Haldane, might realistically be in the range of 10%. In contrast, while I was a plant breeder at Cornell University, I could easily 'spend' (eliminate) 99% of my breeding populations for selection purposes - because of the extreme fertility of plants.

The concept of "cost of selection" is so very important, I need to say more about it. The normal reproductive rate of a species must obviously be at least two offspring for every two adults - or the species quickly goes extinct. However, every species needs much more reproduction than this - just to survive. An *excess population* is needed to "fund" many things - both genetic and non-genetic. For example, there is a huge *random* element to successful reproduction - many individuals in a population die or fail to reproduce for reasons that have nothing to do with genetics. Being hit by a truck or killed in a war has very little to do with a person's "genetic fitness". This cost of random death is absolute, and must be "paid" before we even consider any type of selection. In some populations, this cost (loss), may be 50% of the total population. In such cases, we need at least 4 offspring per 2 adults just to avoid extinction. But at this point, selection has not yet even begun. When we actually consider the genetic factors that determine reproductive success - there are significant genetic features which are *not* passed on to the offspring - which is to say these genetic components are not 'heritable'. For example, many genes work well in certain combinations, but are undesirable all by themselves (this would be true wherever there is *heterosis* or *epistasis*). Selecting for such gene combinations is really "false selection", because it does no good - the gene combinations are broken up in meiosis, and are not passed on to the offspring. Yet such "false selection" must still be paid for - requiring still more reproduction. We have still not started to pay for "real" selection! Real selection can take several forms - stabilizing selection, sexual selection, progressive selection, etc. Each form of selection has a reproductive cost. **All** reproductive costs are additive, and **all costs must be paid for**. Total reproductive costs must never exceed the actual reproductive potential of a species. Only if a species is sufficiently fertile, and there is sufficient surplus population to fund **all other costs**, does any type of selection become feasible. In other words, selection is only possible to the extent that there is residual excess population, after all other costs have first been paid. Selection is a little like discretionary spending for a family on a tight budget. The question always comes down to - "Can we afford it?"

Fitness (due to phenotypic superiority) is actually the real trait that natural selection always acts upon, and this very fundamental trait is actually *very poorly inherited*. This runs counter to popular thinking. According to Kimura, fitness has low *heritability* - even as low as .004 (Kimura, 1983, p.30-31). The concept of heritability is dealt with in more detail in Chapter 6. For now it is sufficient to know that low heritability means that environmental factors are much more important than genetic factors, in determining who appears "superior". For Kimura to say that general fitness has very poor heritability is an amazing acknowledgment! It means that even with intense selection pressure, nearly all of a population's surplus ends up being "spent" to remove non-heritable variations, and thus **most reproductive elimination is unproductive.** In other words selection for general fitness has minimal impact on the makeup of the next generation. Kimura's statement implies that *only a small fraction of the surplus population is* *truly available to pay for the elimination of mutations* (which is exactly what I have been saying). Despite this very important fact, for now I am going to be exceedingly generous - and will assign all available "selection dollars" (surplus population) to the elimination of mutations. Unless otherwise specified, I will always assume that all the surplus population is dedicated exclusively to selection for removal of mutations. However, **the reader needs to understand that in reality, only a very small fraction of any population's surplus can honestly be apportioned to mutation elimination** (see Chapter 6, Figure 8a-c).

I believe one of the most fundamental mistakes that theorists make, as they invent their various scenarios, is to ignore selective cost. They spend their selection dollars like a teenager with a credit card. They speculate as if there is always an infinitely large surplus population. Because theorists are usually unconstrained by realistic cost limits, in their minds they imagine they can "fund" any number of simultaneous selection scenarios. They can spin off one selection scenario, upon another, upon another. This reminds me of the movies - like in the old westerns where the cowboy would fire his "six-shooter" dozens of times without reloading, or like Legalos, in *Lord of the Rings,* who never runs out of arrows! However, such movies are fantasies, and the movie-makers are free to claim "artistic license". Genetic theorists do not have artistic license, and should be held accountable for how they spend their selection dollars - even as an accountant would be held accountable for where the money goes. Theorists should be assigned a realistic number of selection dollars (i.e. some realistic fraction of the surplus population) - based upon the reproductive reality of a given species. They should then "spend" this part of their surplus population soberly, and without any deficit spending.

If this principle were honestly employed, there would be greatly diminished expectations of what selection can really do.

2. Recognizing obscured ("invisible") mutations. If you wish to select against a mutant, you must be able to identify it within the population. Can we identify the carriers of a typical point mutation within the human population? The answer is that we cannot - apart from screening the whole population using very expensive DNA tests. Only those extremely rare point mutations which cause gross physical deformities can normally be identified on a practical level. For most individual point mutations, even though they are destructive and result in loss of information, they are so subtle that they are essentially "invisible" - they do not produce a distinct or recognizable effect. So to artificially select against a typical point mutation, we would need to do expensive lab analyses for every person on the planet. Such extensive testing is at present entirely impractical. So we can see that there are fundamental problems in terms of identifying "good" versus "bad" individuals for selection. Obviously when considering millions of mutations simultaneously, this problem becomes mind-boggling. Imagine wanting to buy an encyclopedia set, knowing that each set of volumes has its own unique collection of thousands of misspellings. Could you realistically stand there in the bookstore and sort through all those volumes - trying to find the set of volumes which was least "degraded"? Given two sets, each of which contains its own unique set of 10,000 misspellings – how would you choose which set has the worst mistakes? The choice would become totally arbitrary! This issue of large numbers of silent or "nearly-neutral" mutations was first recognized by Kimura (1968), and its implications have been explored by Kondrashov (1995).

3. Systematic reproductive elimination. The cost of selection is a huge problem and the "silent" nature of most mutations is a huge problem. The challenge becomes even greater when it comes to preventing mutant individuals from mating. Nowhere on this planet is there a social system which can control human reproduction with high precision. The most infamous case where this was attempted was in Germany under Hitler. That experiment failed catastrophically. Planned Parenthood and modern birth control practices, while effective in reducing average family size, have not been effective in eliminating mutations. In most instances, human mating and reproduction remain essentially random - except for those very rare cases where mutations result in very pronounced genetic defects.

We conclude from our very modest illustration above that **we are not really in a practical position to artificially select for even one point mutation within the human population**. This is sobering. When we then go on to consider multiple mutations, the problems escalate exponentially. Even if we were able to identify all the carriers of numerous mutations and could effectively prevent them all from mating, we would still sooner or later encounter the problem of selective cost - because when we try to select against multiple mutations simultaneously we run into problems of rapidly shrinking population size. So we begin to see that selection is not as easy as we thought! Even the simplest selection requires: 1) maintenance of population size; 2) clear identification of mutants; and 3) effective exclusion of the mutants from the breeding population. As we will see more clearly in the next chapter, when we consider all mutations simultaneously each one of these three requirements becomes utterly impossible.

Natural selection to the rescue? An obvious response to these problems might be - we don't need to do all this - we can let nature do it for us! The problem is that natural selection, like artificial selection, entails exactly the same problems. Natural selection, because of the cost of selection, cannot select against too many mutations simultaneously – or else selection will either become totally ineffective, or will result in rapidly shrinking population sizes (or both). Furthermore, natural selection needs to be able to recognize multitudes of what are essentially "invisible" mutations. Lastly, natural selection needs to be able to somehow exclude multitudes of mutations from the breeding population - simultaneously - which is logistically impossible because of *selection interference*. These very real constraints on natural selection will limit what we can realistically expect natural selection to accomplish.

Genetic selection still works - Please do not misunderstand where I am going with this. I am not saying that selection does not work – on a limited level it certainly does! My career in science involved use of artificial selection as a plant breeder. My colleagues and I were able to regularly breed better plant and animal varieties - which have had fundamental importance to modern agriculture. When I later became involved in genetic engineering of plants - we routinely used selection techniques to recover transgenic (genetically engineered) plants. Likewise, natural selection has eliminated the worst human mutations - otherwise the human race would have degenerated long ago - we would not even be here to discuss all this. But both natural and artificial selection have very limited ranges of operation, and neither has the **omnipotent power** so often ascribed to them. Selection is not a magic wand. While I will enthusiastically agree that selection can shape some specific gene frequencies, I am going to argue that no form of selection can maintain (let alone create!) higher genomes. The simplest way to summarize all this is as follows: **Selection works on the genic level, but fails at the genomic level.**

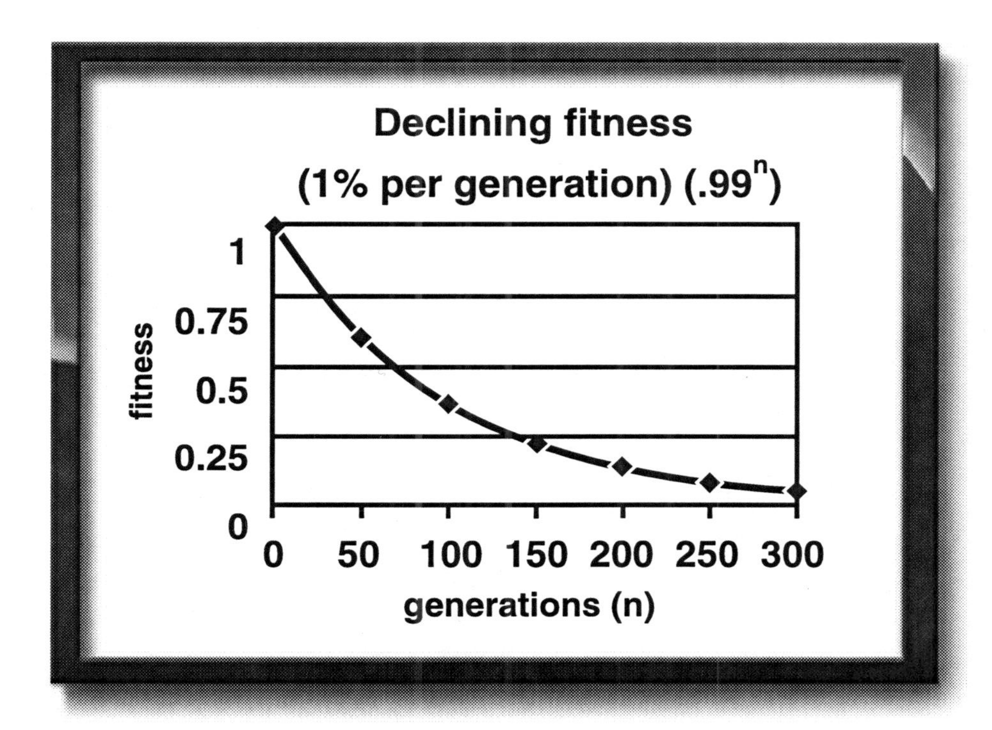

Figure 4.

Dr. Crow (1997) indicates that he believes the fitness of the human race is presently degenerating at 1-2% per generation, due to the accumulation of mutations. A 1% decline in fitness per generation (beginning with a fitness of 1) is shown for a hypothetical human population - over a period of 300 generations (6,000 years). This type of progressive loss of fitness would clearly lead to dramatic degeneration of the human race - within an historical timeframe.

Figure 5.

The Princess and the Nucleotide Paradox. The Primary Axiom requires that natural selection which occurs on the level of an individual (the Princess) must recognize billions of individual nucleotide effects – which only exist on the molecular level (i.e. the Princess' peas). Separating the Princess and the nucleotides is a vast gulf. This gulf is partly separation in scale (to picture this, try to realize that if a nucleotide was as big as a pea, a human body would be roughly 10,000 miles tall.). This gulf is also the separation by level of organization (there are at least a dozen levels of biological organization between a nucleotide and a human being). No nucleotide affects a human body directly – but only through an elaborate labyrinth of nested systems. So the mattresses between the Princess and the nucleotide are many, and they are very thick. To make things still worse, life's self-correcting mechanism (called homeostasis) operates on every biological level – which effectively silences most nucleotide effects – even as modern anti-noise technologies use negative feedback to negate noise. So for our analogy, we would have to incorporate high tech internal "auto-correction" machinery into each mattress. Now we have to add one more dimension to our analogy to make it accurate. The Primary Axiom does not just ask the Princess to sense if there is a single pea beneath her stack of mattresses. Through these mattresses - she must "feel" encyclopedias written in molecule-sized Braille bumps, and decide which volumes have the fewest mistakes!

Can Genomic Selection problems be solved?

Newsflash - Selection cannot rescue the genome.

At the beginning of the last chapter, we began by considering the problems associated with selecting for a single mutation within a human population. Traditionally, geneticists have studied the problem of mutations in this simple way - considering just one mutation at a time. It has then been widely assumed that what works for one mutation can be extended to apply to all mutations. Wow - talk about senseless extrapolation! This is like saying that if I can afford one car, I can afford any number - or if I can juggle 3 balls, I can juggle any number! We are learning that the really tough problems are not seen with single genes or single nucleotides - but arise when we consider all genetic units combined – the whole genome. To understand what is needed to prevent genomic degeneration – we must consider the problem of mutation on the genomic level, and we must implement what I am calling "genomic selection".

There are 3 billion nucleotide positions (each with 2 copies) in the genome, and so there are 6 billion possible point mutations (misspellings). The profound difficulties of making mutation/ selection work on the genomic level have only been recognized by a few far-sighted geneticists in the past (Haldane, Muller, Kimura,

Kondrashov) - but the whole sticky problem has repeatedly been swept under the rug. This is because it creates insurmountable problems for evolutionary theory. In the last few decades, we have learned that the genome is much larger and more complex than anyone could have imagined. We have learned that the human mutation rate is much higher than was previously thought possible. We are learning that the actual percentage of mutations that are truly "neutral" is steadily shrinking, and the percentage that actually *add* information (if any) is vanishingly small. For all these reasons, we cannot ignore the problem of genomic degeneration any longer. We must ask - "Can *genomic selection* solve the problem?"

The bottom line is that virtually every one of us is exceedingly *mutant*. This destructive mutation process has been going on for a long time. Therefore, in addition to the roughly 100 new mutations we each have added to the human gene pool (our personal contribution to the problem), we have inherited a multitude of mutations from our progenitors. So to put the problem of genomic selection in proper perspective, we have to take the selection problem posed at the beginning of the last chapter (for a single point mutation at a single nucleotide site), and then multiply this problem by a factor of <u>at least one billion</u>. *(I am allowing that 2/3 of all nucleotide positions are "junk DNA" and may be truly neutral - and thus would not need selection. However, it is not clear that any part of the genome is truly neutral).* Can you start to see that selection against mutations on the genomic level is fundamentally different from our first illustration - where we were just selecting against a single mutation?

1. Cost of selection - The fact that all people are mutant makes selection much more difficult. If we were to simply select against all "mutations", it would mean no one could reproduce - resulting in instant extinction! Obviously this selection strategy creates a "reproductive cost" that is too high! It is widely acknowledged that we each inherit thousands of deleterious mutations - and so collectively as a population we carry many *trillions* of deleterious mutations. However, to make the problem easier, let's limit our attention to just the 600 billion new mutations that entered the human gene pool within our own generation. Since we cannot simply select against "mutants" - we will have to select between individuals who are "more mutant" versus those who are "less mutant". (As we will see, recognizing "more mutant versus less mutant" is a huge problem in itself.) All this selection must cost us considerably less than 33% of the population per generation.

Let me try to illustrate the extent of the cost problem which is associated with selecting against 600 billion mutations. If we have a population of 6 billion people, then maximally only one third of them could be "eliminated" (i.e. prevented from having children). This is 2 billion people (try to imagine that - this thought should be enough to make even the most cold-blooded eugenicist shudder). Yet what good would this Draconian measure accomplish? Preventing 2 billion people from mating would only eliminate 100 x 2 billion = 200 billion new mutations. This would still leave 400 billion new mutations as the newly added *genetic burden* for the next generation! Even if we assume that two-thirds of the remaining mutations are perfectly neutral, we still have 133 billion deleterious mutations added to the population. We just can't get rid of enough of the mutations and still maintain population size! Even if two-thirds of the mutations are neutral, and in addition we doubled selection intensity (although we certainly cannot really afford to spend two-thirds of our population), it would still leave 67 billion new deleterious mutations for the next generation. The cost of selection clearly limits how many mutations we can eliminate per generation, and the known mutation rate for humans is way too high to be countered by any level of selection. **Therefore, mutations will continue to accumulate, and the species must degenerate!** Can you see that the cost of selection is rather a mind-boggling problem when viewed on the genomic level?

2. Obscured or "invisible" mutations - Surprisingly, when it comes to selection, lethal and near-lethal mutations are not the real problem - at least not from the whole population's point of view. Such mutations are rare and are self-eliminating. Likewise, absolutely neutral mutations do not really matter (if they exist). It is those mutations of "minor" effect that really do the most damage - especially within the short time frame (Kimura and Ohta, 1971, p.53). Selection must prevent the accumulation of minor mutations, or the species will rapidly deteriorate and fitness will decline. However, as we will see, even if selection could keep minor mutations in check (it can't), it appears to be powerless to stop the accumulation of the most abundant class of mutations - which are called *nearly-neutral.* Therefore, in the longer run, higher genomes will all eventually degenerate - with or without selection.

a) Nearly neutral mutations -

Nearly-neutral mutations have infinitesimally small effects on the genome as a whole. Mutations at all near-neutral nucleotide positions are automatically subject to random drift – which means that they are essentially immune to selection. Their fitness effects are so miniscule that they are masked and overridden by even the slightest fluctuations, or *noise*, in the biological system (Kimura, 1968; 1983); and Kimura and Ohta, (1971). These are the most abundant mutations - as shown in the "near-neutral box" in Figure 3d.

Since most mutations should be near-neutral, and since they are so subtle that they cannot be selected, why are they important? They matter because those nucleotide sites contain information, and their mutation results in the gradual but certain erosion of information. **Collectively, near-neutral nucleotides must account for most of the information in the genome.** This is just as true as the fact that all the seemingly insignificant letters in this book can collectively add up to my clear message. If we start with a very long and complex written message (i.e. an encyclopedia), and we start to introduce typographical errors, most of the individual errors will only have an extremely trivial effect on the total message. Individually, they are truly insignificant. But if this process is not halted, eventually the message will become corrupted, and it will eventually be completely lost. An alternative example would be the rusting of a car. As our car ages, we can repair the big stuff - replace tires, fix dents (like selection for major and minor mutations), but we cannot stop the rusting process, which is happening one atom at a time (near-neutral mutations). Each iron atom that oxidizes seems perfectly insignificant, but added up across the entire car, the process is certain and deadly. A third example would be the aging of our bodies. We can repair teeth, do facelifts, even replace hearts. But it is the cumulative aging of the individual cells (principally due to mutations) which places a specific limitation on our lifespan. This is true even though each individual cell is trivial and entirely expendable. Just as the human body *rusts out* due to countless microscopic mistakes (all of which in themselves are insignificant), the human genome must also be "rusting out" due to near-neutral mutations. No selection scheme can stop this process. This is the essence of the near-neutral mutation problem. This whole problem has led one prominent population geneticist to write a paper entitled

– "Why aren't we dead 100 times over?" (Kondrashov, 1995). The problem of near-neutrals being un-selectable is very real.

A very large and homogeneous population, within a homogenous environment (for example – a typical bacterium) is more resistant to genetic drift, because it sees much less noise, and experiences much more efficient selection. Such populations usually have simpler genomes, fewer mutations per genome, and far fewer inter-genic interactions. Furthermore, they exist in large numbers, and have very high rates of reproduction. Perhaps most importantly, *every cell is subject to selection, independently, every cell division.* For all these reasons, selection in such systems is much more effective, precise, and can have much higher resolution. This means that in bacteria, a much smaller proportion of the genome is near-neutral and un-selectable (this is why theorists typically prefer to use microbial examples).

Unfortunately, mammals such as ourselves have none of the advantages listed above. We are very much subject to high levels of reproductive noise, we have a large genome, high mutation rates, high levels of gene interaction, and we have very serious constraints on selection. This is why in man, the proportion of mutations which are virtually "un-selectable" should be very large, and the change frequency of such mutations within the population should be entirely controlled by random genetic drift. All such nucleotide positions will mutate freely, and all information encoded by them will degenerate over time.

b) Selection threshold for too many minor mutations - Minor mutations, by definition, have a small but distinct effect on reproductive potential. These are the mutations immediately to the left of the near-neutral box, in Figure 3d. While artificial breeders may have a hard time seeing these subtle changes, natural selection (which is really just another way of saying differential reproduction) can generally "see" them - since by definition they affect reproductive probability. Furthermore, the effects of such mutations are partly additive, so natural selection can select for *numerous* minor mutants simultaneously. In fact, the way natural selection works is very elegant and appears to be designed to stabilize life - which would otherwise very quickly deteriorate. It is really a very wonderfully designed system.

However, selection for minor mutations has significant limitations. The orderly elimination of minor mutations is seriously disrupted by what I call noise. Natural selection must *see* the mutants in terms of the mutants being a significant factor in reproductive probabilities. Even "Mother Nature" can have trouble seeing minor mutations, when other factors are significantly disturbing reproductive probabilities. This is because the differences in reproductive fitness caused by minor mutations are very subtle, and the effects of other factors can be very large. It is a little like trying to see the ripples produced by a pebble when thrown into a storm-tossed sea.

All other variables which affect reproduction, when combined, will significantly interfere with natural selection against any given minor mutant. For example, a high rate of accidental deaths in a population will override and obscure the subtle effects of minor mutations. Likewise, selection for lethal and near-lethal mutations (which must automatically take precedence) will override the more subtle effects of minor mutations. The fact that most mutations are recessive dramatically masks their negative fitness effects, and greatly hinders selection against them. Likewise, all interactions between genes (called epistasis) will interfere with selective elimination of minor mutations. In smaller populations, the randomness of sexual recombination (chromosome-segregations and gamete-unions are both random and thus fluctuate) can routinely over-ride selection. All these effects cause the fundamental phenomenon of genetic drift. Genetic drift has been extensively studied and it is well known that in small populations, it can override selection against all but the most severe of mutations. Breeders, such as myself, know that *all* extraneous effects on reproduction will interfere with effective selection. So a nucleotide's abundance in a population will tend to drift randomly and become immune to selection whenever the net affect of all other factors combined has a greater effect on reproductive probability than does the nucleotide itself.

Again, trying to put the issue in more familiar terms - selection for very subtle genetic effects is like trying to hear a whisper in a noisy room. The softer the whisper, or the more complex the message, or the louder the background noise - any of these contribute to the loss of the message. Selection against a minor mutation works best when its fitness effect is still moderately 'loud', and where there is minimal biological 'noise'. Despite all these very significant limitations, selection for a number of minor mutations still works. Thank goodness it works on this level - otherwise we would not be here!

While selection can definitely work for numerous minor mutations, as the **number** of those mutants increases, each mutant's fitness-effect becomes less and less significant in terms of total reproductive probability. As the number of minor mutations increases - the individual mutation effects become less and less significant, and the efficacy of selection for each one moves toward zero. This problem is alluded to by Kimura (1983). I have demonstrated this problem mathematically in Appendix 2, and the results are shown in Figures 6a-c. Each time we add another trait that needs to be selected for, the maximum selective pressure that can be applied to each trait individually must decline. As the number of traits undergoing selection increases, selection efficiency for each trait rapidly approaches zero, and the time to achieve any selective goal approaches infinity. According to my calculations (see Appendix 2), for a population such as our own, the maximal number of mutations which could be selected simultaneously is approximately 700. Kimura (1983, p.30) alludes to the same problem, and although he does not show his calculations he states that only 138 sites can undergo selection simultaneously, even for a population with very intense total selection pressure (50% elimination), and very weak selective elimination per trait (c = 0.01). Trying to select simultaneously against more than several hundred mutations should clearly lead to cessation of selective progress. Yet even in a small human population, millions of new mutations are arising every generation, and must be eliminated! In the big picture, we really need to be selecting against billions, not hundreds, of mutations. Even in the very limited case of selecting for just a few hundred genetic loci, although it is theoretically possible to do this, it is noteworthy to point out that such highly diluted selection per trait greatly affects the rate of selective progress – which essentially grinds to a standstill. As the number of loci under selection increases, the rate of selective progress (per trait) slows very rapidly, approaching zero. The resulting rate of genetic change would be glacial at best - requiring hundreds of thousands of generations of selection to significantly affect even this very limited number of nucleotide positions.

In a sense, as we select for more and more minor mutations, each mutation becomes noise for the others. At a distance, a room full of whisperers is full of noise, and devoid of net information. As each mutation's effect becomes less and less significant, its individual whisper gets softer, and so the problem of overriding noise gets worse. Soon, even under the best of selection conditions, each individual whispered message cannot be discerned. The effect of any given mutant becomes totally insignificant in light of all other reproductive factors. At this threshold point, the mutation becomes effectively neutral, and all selection abruptly ceases. As we select for more and more minor mutations, we must always reach a threshold point where selection should largely break down. Above a certain number, all minor mutations should become un-selectable - even when conditions are ideal and noise is minimal. Simultaneous selection against too many minor mutations should lead to zero selective progress, and genetic drift takes over. In essence, selecting for too many minor mutations simultaneously, effectively makes them all behave as *near-neutrals* as described in the section above.

Haldane (1957) and Kimura (1983, p.26) both agree that it is not possible to select for a large number of traits simultaneously - due to the cost of selection, as I have described above. This simple reality makes "genomic selection", involving millions of nucleotides, virtually impossible.

3. Reproductive elimination -

We have learned that we cannot stop genomic degeneration because of the high number of mutations occurring in the human population, and the prohibitive reproductive cost of eliminating each one. If that was not enough of a problem, we have learned that we cannot stop genomic degeneration because most mutations are near-neutral - so their effects are obscured and essentially undetectable above biological noise. This makes them immune to selection and subject only to drift and degeneration. Furthermore, we have learned that if we try to select against too many minor mutations simultaneously, they effectively all become like near-neutrals – so they also become un-selectable and subject to random drift. Lastly, I would like to argue that we cannot stop genetic degeneration because we cannot effectively enforce the reproductive elimination of large numbers of mutants simultaneously, for logistical reasons. I have called this problem *selection interference.* This problem has not been addressed sufficiently, but has simply been recognized as a factor interfering with selection (Haldane, 1957; Lynch, Conery, and Burger, 1995; Kondrashov, 1995). **When attempting simultaneous selection for tens of thousands - or millions - of different mutants in the genome, the problem of selection interference becomes absolutely overwhelming.**

Selection interference occurs when selection for one trait interferes with selection for another trait. For example, a desirable trait will routinely be found along with an undesirable trait, within the same individual. To select against the *undesirable* trait automatically means that you are also unintentionally selecting against the associated *desirable* trait (we have to accept or reject the whole person). This association between traits can be very tight (both traits coded for by the very same gene, or two genes side-by-side on the same chromosome), or the association can be loose (simply two genes in the same individual). Even if mutations are only loosely associated in the same individual, for that single generation the two traits are still "linked". Any mutant must always be temporarily linked to thousands of other mutants, in every individual and in every generation. Therefore, selection can never operate on any given mutation in isolation. To select for any given beneficial mutation will always automatically multiply a host of associated deleterious mutations. This problem is inescapable.

To illustrate this problem, let us imagine selecting between two individuals in a population. Because each individual's genes are drawn from the same "gene pool", any two individuals will on average have about the same number of mutations, which will have approximately the same net deleterious effect. When contrasting two such individuals (to discover who is more "fit"), we may find that each has roughly 10,000 different mutations – so there are 20,000 differences between the two individuals. Now when contrasting them, each will have about 10,000 "bad" genic units, and also about 10,000 "good" genic units (the "good" units are the non-mutant nucleotides, corresponding to the other individual's mutations). Due to averaging, the actual difference in genetic fitness between them will be small - and will hinge on just a few *major impact* nucleotide differences. Who will actually be reproductively favored? Because of their high degree of overall similarity in genetic fitness, reproductive success will depend more on random chance and noise factors, than on true genetic fitness. But even if the "better" individual actually is favored in reproduction - almost no selective progress will be made. The individual that is favored will have, essentially, just as many mutations as the rejected individual. We will have *selected away* the 10,000 mutations in one individual, but at the same time we will have *multiplied* another 10,000 mutations in the other individual. Almost all selection is canceled out - 10,000 steps forward, and 10,000 steps backward. The only net gain is for those few 'major' genes which actually made the real difference in fitness. **As we keep seeing, in higher genomes selection can only be effective for a limited number of "significant" nucleotide differences**. The vast bulk of mutations, which will be minor or near-neutral, will cancel each other out and be "un-selectable". On the genomic level, even if we could have perfect control over the reproduction of individuals, we would still fail to effectively prevent the propagation of the vast bulk of delitionary mutations. This problem, to my knowledge, has not been adequately addressed by others – although it is often alluded to by population geneticists.

Selection interference due to physical linkage - The most obvious and extreme form of selection interference is when there is tight physical linkage between beneficial and deleterious mutations. This results in an irreconcilable problem which has been termed *Muller's Ratchet.* Essentially all of the genome exists in large linkage blocks (Tishkoff and Verrelli, 2003), so this problem applies to virtually every single "building block" of the genome. If we refer back to Figure 3d, we can see that mutations are overwhelmingly deleterious - but there should be a few extremely rare beneficial mutations. These very rare beneficial mutations might seem to leave a very slight glimmer of hope for forward evolution. This would not be a rational hope – because such beneficials will overwhelmingly be nearly-neutral and thus un-selectable, and because the accumulating deleterious mutations will always outweigh the beneficials. Yet as long as those rare beneficials are on that graph – they seem to offer a glimmer of hope - to the hopeful. The problem of physical linkage erases those beneficials from our graph (Figure 7). This should completely eliminate any trace of rational hope for forward evolution.

Within any given physical linkage unit, there should be, on average, thousands of deleterious mutations accumulated before the first beneficial mutation would even arise. Therefore, there would never arise at any time even a single linkage group within the whole genome which could realistically experience a net gain of information. Every single beneficial mutation would always be inseparably tied to a large number of deleterious mutations. This can be visualized graphically (Figure 7). In Figure 3d, we mapped the distribution of the effects of single mutations. We can do the same thing in terms of the mutational effects of linked mutation clusters. Because these clusters never break apart, the net effect of any cluster of mutations will be inherited as if it were a single mutation, and the effect of any mutation cluster would simply be the net affect of all its component mutations. By the time that at least two mutations per linkage block have accumulated, nearly every beneficial mutation will have been canceled out by a linked deleterious mutation. At this point the distribution will already show essentially zero linkage blocks with a net gain of information (see Figure 7). To illustrate this point further, if only one mutation in a million is beneficial, then the probability of a linked pair of mutations both having a net beneficial effect becomes too small to even consider (10^{12}). As more time passes, the average number of mutations per linkage group will increase – such that the net loss of information per linkage group will increase, and the complete disappearance of net-gain linkage groups will very rapidly approach *absolute certainly*. The human genome is a composite of roughly 100,000-200,000 linkage blocks. Based upon the logic provided above, we can know with very high certainty that every single one of these "building blocks of evolution" is deteriorating.

Based upon numerous independent lines of evidence, we are forced to conclude that the problem of human genomic degeneration is real. While selection is essential for slowing down degeneration, no form of selection can actually halt it. I do not relish this thought, any more than I relish the thought that all people must die. **The extinction of the human genome appears to be just as certain and deterministic as the extinction of stars, the death of organisms, and the heat death of the universe.**

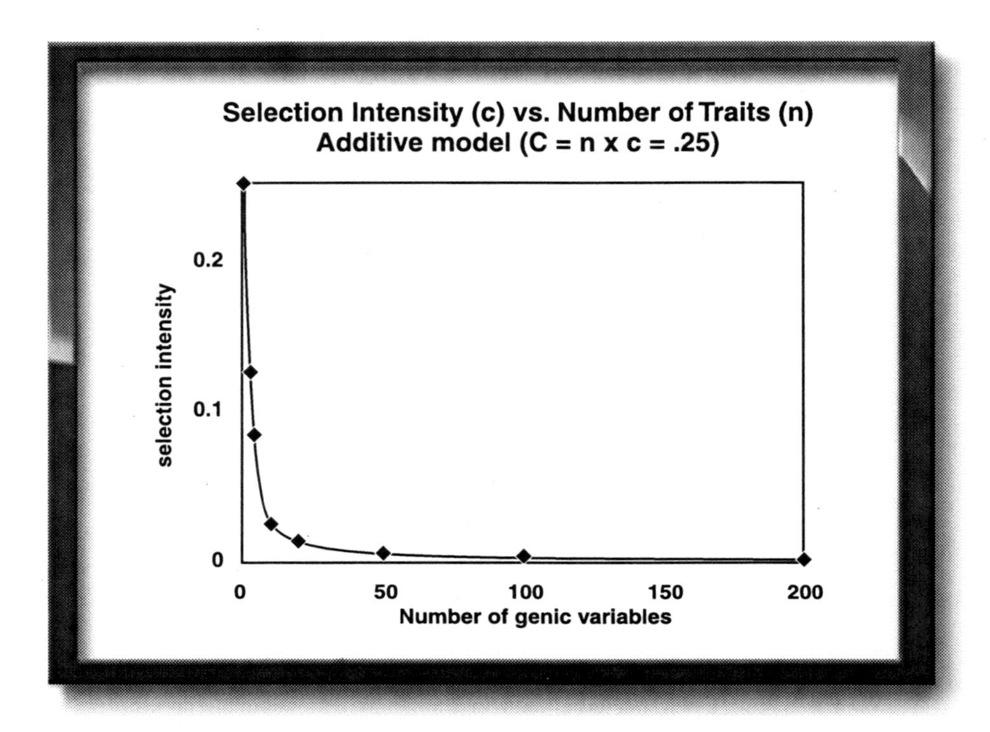

Figure 6a.

Selection Threshold - Additive Model. Selection always involves a reproductive cost (C), meaning that some individuals cannot reproduce. Total selection cost must be substantially less than a species' excess reproduction, or the population will rapidly shrink and face extinction. As more traits are under selection (n) , the total cost attributable to each trait (c) must diminish rapidly, so that the total cost does not exceed the population reproductive potential. To visualize the problem of selection threshold, I have plotted maximal allowable selection intensity per trait (c) against number of traits under selection (n), for a population which can afford to lose 25% of its individuals for elimination of mutations (C=.25). As can be seen, the allowable selection pressure per trait plummets extremely rapidly as number of traits increases. Selection pressures will obviously approach zero very rapidly, and a threshold point will be reached where each trait is effectively neutral. This curve is based on an additive model, following the formula: C = n x c (see Appendix 2).

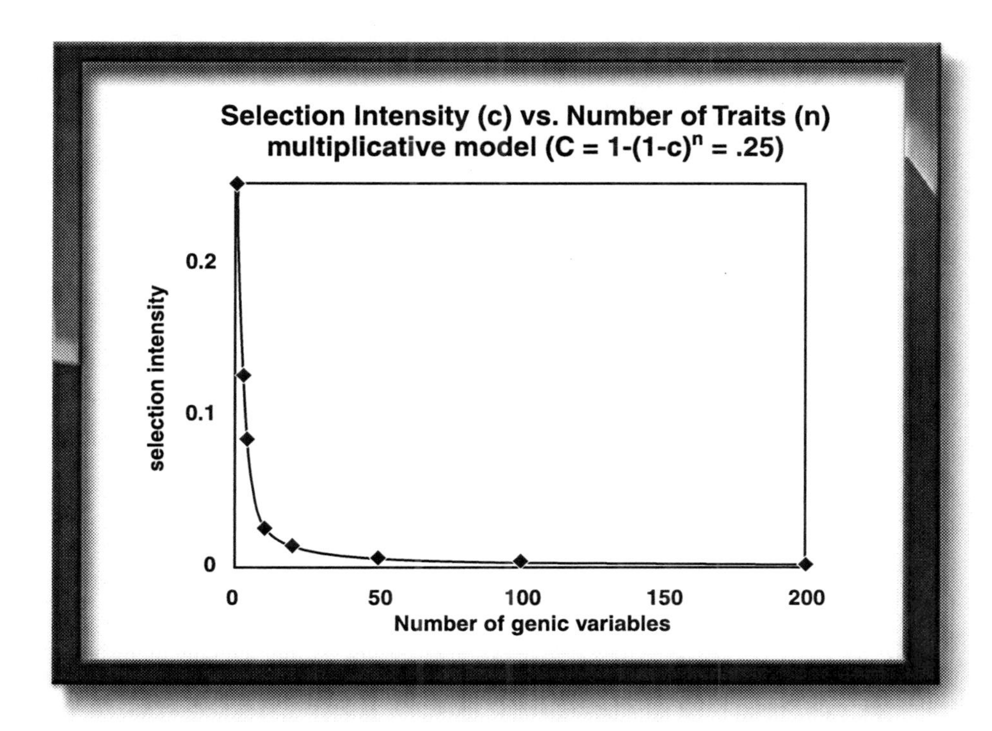

Figure 6b.

Selection Threshold - Multiplicative Model. The additive model illustrated in Figure 6a can be contrasted to a multiplicative model. The curve above is based on a multiplicative model, following the formula: $C = 1 - (1\text{-}c)^n$ (see Appendix 2 for more detail). The two curves in 6a and 6b are essentially identical. In both cases, maximal allowable selection pressures rapidly approach zero, as number of traits under selection increases. In both cases a threshold point will rapidly be reached where each trait is effectively neutral.

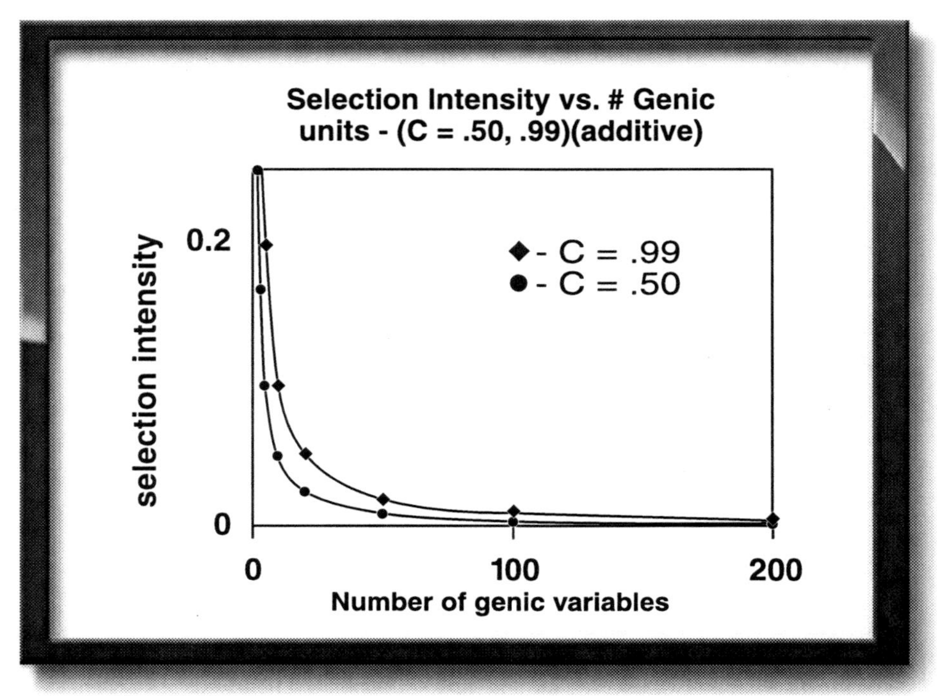

Figure 6c.

Selection threshold for extremely fertile populations. Doubling human fertility, up to very unrealistic levels (C=.5), does not significantly reduce the problem shown in Figure 6a. Even for extremely fertile species, such as plants, where C may be .99 (100 offspring per plant), the problem of selection threshold is still very real. Selecting for many traits simultaneously decreases selection efficiency for each individual trait, until selection reaches a point where it is entirely ineffective. These curves follow the additive formula: C = n x c.

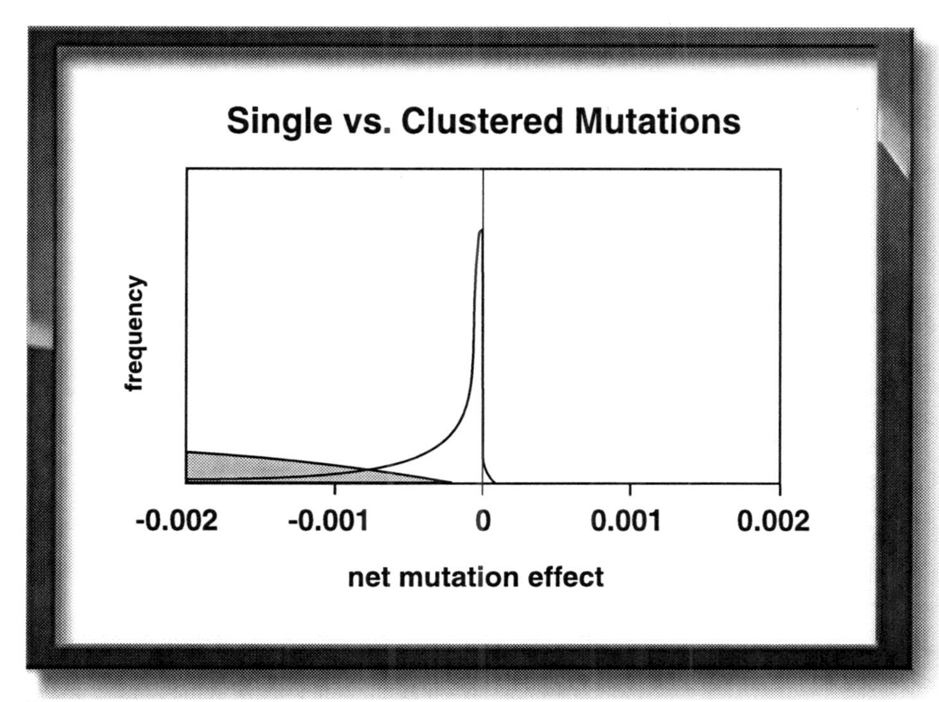

Figure 7.

Mutation clusters will always be 100% deleterious. In Figure 3d, we saw that the distribution of individual mutations is tightly crowded toward the neutral mark (the steep curve straddling zero), and that while there are many deleterious mutations, there are vanishingly-few beneficial mutations. These rare beneficial units completely disappear when we analyze mutations as they actually occur - within clusters which are physically linked and inherited as if a single trait. The actual distribution of the effects of linked mutation clusters (the shaded distribution curve), will be strongly shifted to the left, when compared to individual mutations. Any linked mutation cluster will be inherited as a single genetic unit, and the effect of that mutation cluster will simply be the net effect of all of its component mutations. The fitness effect of any cluster can be calculated to be the *average effect of its individual mutations times the number of mutations within that cluster*. Any rare beneficial will be quickly cancelled out – with high certainty. Since the vast majority of mutations are deleterious, each mutation cluster will have an increasingly negative affect on fitness, each generation. By the time there are just two mutations per linkage group, nearly all beneficial mutations will have been cancelled out by at least one linked deleterious mutation. As the mutations accumulate beyond two mutations per cluster, it becomes increasingly certain that there will be no linked cluster left with a net beneficial effect.

Chapter 6

A closer look at noise.

Newsflash - The problems are not really so bad – they're much worse!

If you want to receive a message on the radio, you need to limit the amount of "noise", or static. If there is a strong interfering signal - the message is lost. It is for this reason that governments can "jam" radio programs which they do not want their people to hear. Likewise, if governments do not regulate the allowable band-width for each radio station, stations very soon become noise *for each other*. Radio static can arise from many sources – other radio stations, solar flares, cosmic radiation, radio waves being reflected by cloud cover, electric motors that are running in the neighborhood, local walkie-talkies, etc. Regardless of the source, noise results in loss of information.

A very weak signal is easily destroyed by any amount of noise. The weaker the signal and the greater the noise, the more certain is that loss of information. A low signal-to-noise ratio always ensures loss of information. When we receive a signal plus noise, *amplification* does not help us. Turning up the volume on your radio does not help overcome static. We just wind up amplifying the static as much as the signal. To ensure minimal loss of information, there must be a favorable signal-to-noise ratio.

The reason that most nucleotides must be un-selectable is because of consistently low signal-to-noise ratios. Likewise, the reason we cannot select for many nucleotides simultaneously is because of rapidly shrinking signal-to-noise ratios. In fact, one of the primary reasons that selection cannot save the genome is because of ubiquitous "noise". When we apply selection to the entire genome, the signal-to-noise ratio quickly approaches zero. Hence in the big picture, noise will consistently outweigh the effects of individual nucleotides. ***This is a primary reason why selection works on the level of the gene, but fails on the level of the genome.***

Heritability - the important difference between genotype and phenotype - In genetics, the signal-to-noise ratio is often expressed in terms of "heritability". If a trait has high heritability, this means that most of the variation observed for that trait is genetically heritable, and so this trait will be easy to select for. The essence of this concept of heritability is simply the ratio of heritable versus non-heritable variation. Non-heritable variation is largely due to variation within an organism's individual environment, and is the source of "phenotypic noise". So a genetic heritability value for a trait is essentially equivalent to a signal-to-noise ratio. For example, any observed difference in the intelligence of two people will be partly due to heritable genetic differences (genic differences which can be passed from parent to child), and partly due to environment (i.e nutrition, the quality of training, etc.). So intelligence is determined partly by *nature* (inherited capabilities), and partly by *nurture* (non-inherited characteristics). This is equally true for height, speed, weight, etc. So in a sense heritability can be understood as a reflection of the ratio of *nature* versus *nurture*. When heritability is "1" for a trait - that trait is 100% hereditary (i.e. blood type), and it is not affected by environment at all. If heritability is "0" for a trait - that trait is not inherited at all, it is entirely environmental in nature (i.e. a tattoo). A simple trait such as height is usually quite highly heritable ($h^2 = 0.3$). This specifically means that for such a trait, 30% of the phenotypic variation is heritable (selectable) variation. Unfortunately, for very complex traits such as "fitness", heritability values are low (i.e. .004) – often approaching zero (Kimura, 1983, pp.30-31). This is because *total fitness* combines all the different types of *noise* from all the different aspects of the individual.

When Kimura says that fitness heritability is generally very low (i.e. .004), he means that almost all variation for individual fitness in due to non-heritable (non-selectable) effects. This means that almost all selection for fitness will be ineffective and *wasted*. **Low heritability means that selecting away bad *phenotypes* does very little to actually eliminate bad *genotypes*.** To illustrate this, consider seeds from a cottonwood tree. Some seeds will land on fertile ground ideal for growth, with abundant moisture, few weeds, etc. But most seeds will land on places that are too dry, too wet, too many weeds, too much shade, too many people, etc. The result will be great diversity in the health and vigor of the resulting trees, and huge differences in their survival and reproduction. But almost all of this "natural selection for the fittest", will really only be selection for the *luckiest* - not the genetically superior. Hence in most natural situations, most phenotypic variation in fitness will only be due to non-genetic noise, and will have very little to do with heritable differences. This is what we mean by low heritability. Low heritability largely neutralizes the effectiveness of selection. Like an automobile with a racing motor but a broken transmission, there can be lots of selection happening, yet almost no genetic progress.

Let's consider in more detail what is involved in non-heritable variation for fitness, by referring to Figure 8a. Obviously, variation in environment can create major differences between individual phenotypes. It is often estimated that about 50% of all phenotypic variation is due to just environmental variation. If one plant is growing better than another plant in the same field, there is a high probability it is growing better just because it sits on a slightly more favorable piece of soil. This aspect of phenotypic variation is shown graphically as sector #1 on our pie-chart. This type of variation strongly interferes with effective selection, by adding to non-heritable noise and diminishes the signal-to-noise ratio.

The second largest part of phenotypic variation is what is called environment-by-genotype interaction. This is often estimated to represent about 25% of all phenotypic variation. Given our two plants in the field, if we change the environment by irrigating the field, the extra water may be good for one plant but bad for the other – depending on their genotype. This type of variation, although it has a genetic component, is not consistently heritable. Like environmental variation, this aspect of phenotypic variation just adds to noise and interferes with selection. This is shown as sector #2 in our pie-chart.

The third largest part of phenotypic variation is non-heritable genetic variation. That may sound like a contradiction, but it is not. A large part of genetic variation is due to factors that are not passed down consistently from generation to generation. These factors include epigenetic effects (sector 3), epistatic effects (4), dominance effects (5), and genetic effects subject to cyclic selection (6). To make a long story short, most genetic variation is not heritable – at least not in a linear and selectable manner.

The only fraction of genetic variation that is heritable (and therefore potentially selectable) is what is called additive genetic variation (sector 7). Very simply, additive genetic variation is where a given trait (or nucleotide) is unambiguously and consistently better than an alternative trait (nucleotide) within the population. For a complex trait such as fitness, this type of additive genetic variation makes up a very small part of the total phenotypic variation. If Kimura's estimate is correct that fitness typically has a heritability of only about .004, then only about 0.4% of phenotypic variation for fitness is selectable. This represents a signal-to-noise ratio of about 1:250. One way of expressing this is that 99.6% of phenotypic selection for fitness will be entirely *wasted.* This explains why simple selection for total phenotypic fitness can result in almost no genetic gain.

If we have a trait such as fitness which has low heritability (Figure 8a), and we have a species of low fertility such as man (Figure 8b), we can see that only a tiny part of a population can be used for effective selection (Figure 8c). If we can only selectively eliminate about 16.7% of a population, and only 0.4% of that selection is actually effective, then only 0.07% of that population can be employed for truly *effective selective* elimination. In other words, less than 1 person in 1,000 is available for the *effective* elimination of all deleterious mutations, and for the *effective* fixation of any possible beneficial mutations.

The heritability for a single *trait* such as total fitness can be remarkably small, yet the 'heritability' of a typical *nucleotide* is infinitesimally smaller. Let us consider the "heritability" of an average single nucleotide mutation (this is an unorthodox but useful application of the concept of heritability). The "signal" (i.e. the heritable additive fitness value of such a nucleotide) is inherently too small to measure. But the "noise" is astronomical – it is the effects of all the non-heritable components of variation, **plus the effects of all the other segregating nucleotide positions!** In a typical population there are millions of other segregating nucleotides. So the effect of an average single nucleotide will very consistently be lost in an *ocean of noise*, with signal-to-noise ratios consistently less than one to a million. The heritability of such a nucleotide is not significantly different from zero – explaining why most nucleotides are inherently un-selectable and must be termed "nearly-neutral" by Kimura's definition.

Another major source of noise – probability selection, not threshold selection As a plant breeder I would score hundreds of plants for their phenotype (yield, vigor, disease resistance, etc), and then I would rank them - from best to worst for overall performance. I would then decide what fraction of the population I wished to eliminate, drawing a line through the ranking at the desired level. Every plant above the mark was kept for mating, every plant below the mark was destroyed. This is called "truncation selection". This is a very artificial type of selection, and is used by breeders because it is especially effective. However, this type of selection never happens in nature. Natural selection is always based only upon probability. Mother Nature does not tabulate for each member of a population some fictional "total fitness value" based upon total phenotypic performance for all traits combined. Mother Nature does not then rank all the individuals. Lastly, Mother Nature does not draw an arbitrary line, and eliminate all individuals below that line. Instead, the phenotypically inferior individuals simply have a slightly lower probability of reproduction than the others. Very often, by chance, the inferior individual will reproduce, and the superior individual will not. In fact, there is only some modest correlation coefficient, which relates phenotypic superiority with reproductive success. The more reproductive "noise" (i.e. random survival/mating), the weaker the correlation becomes, and the less certain it becomes that a superior individual will actually be favored in reproduction. If we realistically disqualify all types of truncation selection from the evolutionary model, the result is the addition of a whole new level of noise, which further reduces the effectiveness of selection.

The nature of the non-truncation problem is easy to illustrate. Picture a population of shrimp swimming en masse, and a whale comes and swallows half the population in one bite. Is this an example of "survival of the fittest" among the shrimp? Was there a precise ranking of most fit to least fit, followed by a strict cut-off type of selection? Alternatively, picture a large mass of frog eggs in a river. A large number of eggs are eaten by fish, even before they hatch. A large number of tadpoles are picked off by birds, a boat is launched and squishes a few hundred more, many more are swept over a waterfall. Many maturing adults burrow into a mud bank which is later removed by a dredging operation, and most of the surviving adults are picked off by more predators. Almost all the elimination has been random. Once again - we are seeing survival of the *luckiest*. So where is the systematic sorting by phenotype? It is largely absent! There is a huge element of noise - not just in determining who has the best phenotype - but also in terms of differential survival and reproduction which is *in spite of phenotype*. This noise, which affects reproductive success *in spite of phenotype,* is noise which is *over and above* the noise we have considered in the heritability section above. Therefore we should never model truncation selection in any honest natural selection scenario. Furthermore, because of high levels of reproductive noise, we must honestly use a relatively low correlation coefficient when calculating the probability that a superior phenotype will reproduce. **Perhaps 50% of reproductive failure is independent of phenotype (Figure 8b).**

The third level of noise – gametic sampling - There is a third level of genetic noise. This is the statistical variation associated with small population sizes. If you toss a coin many times, you will predictably get about 50% heads and 50% tails. However, if you toss the coin just 10 times, there is a good chance you will *not* get 50/50. You may get 60/40 or 70/30, or you might even get 100% heads. The frequency of each possible outcome is readily predicted by using probability charts. This same type of statistical variation occurs when a gene or nucleotide is segregating within a population. In very large populations, gene segregations tend to be highly predictable, but in smaller populations gene frequency will fluctuate very significantly, and in a random manner – just like a series of coin tosses. Such statistical fluctuations result in what is called genetic drift – which means gene frequencies can change - regardless of the presence or absence of selection. This simple probability element of fluctuating gene frequencies is well studied.

Classically, population geneticists have dealt with genetic noise only on the level of this last type of noise - probability fluctuations. This particular type of genetic noise, which I am going to call *gametic sampling*, is very sensitive to population size. In small populations random genetic drift is very strong and can override the effects of even substantial mutations. This is why the small populations of endangered species are especially subject to what is called *mutational meltdown.* In small populations, natural selection is largely suspended. It is on the basis of this type of noise (from just gametic sampling), that Kimuru first defined his "near-neutral" mutations. It is for this same reason that Kimura calculated the size of his "no-selection box" (Figure 3d) as a simple function of population size (either plus or minus $1/2N_e$). It is very attractive for the genetic theorist to limit consideration of noise to just gametic sampling. This is because one can conveniently make noise from gamete sampling largely disappear, simply by imagining larger populations. At the same time the theorist can make selection itself periodically "go away" by invoking high-noise episodes associated with small population *bottlenecks* (i.e. the Out-of-Africa theory).

I believe it is extremely important to realize that gametic sampling is only a minor part of total genetic noise, and that the other two important aspects of genetic noise (as discussed above) *are only partially diminished in large populations.* So we can never really make noise "go away" by invoking larger population sizes. Noise is *always* present, and at much higher levels than is normally acknowledged by population geneticists - even within large populations. In fact, very large populations invariably will have *enhanced* noise. This is in part because of population substructure (many smaller sub-populations – each with its own gametic sampling fluctuations). It is also because bigger populations extend over a wider range of environments, becoming subject to even more environmental variations. Large population size does not reduce the random elements of reproduction, nor does it reduce the phenotypic masking of genotype. In terms of natural selection, noise remains a severe constraint <u>always</u>. Under artificial conditions, plant and animal breeders have been able to very successfully select for a limited number of traits. They have done this by employing intelligent design in their experiments, to deliberately minimize noise. They have used blocking techniques, replication, statistical analysis, truncation selection, and highly controlled environments. Natural selection does none of this - it is by definition a blind and uncontrolled process – subject to unconstrained noise and unlimited random fluctuations.

The genetic consequences of all this noise - When we realize that high levels of genetic noise are unavoidable, we realize we cannot wave away the near-neutral box in Figure 3d – not even by invoking larger population sizes. As we come to appreciate that there are two entire levels of noise above and beyond gametic sampling, we realize that Figure 3d must be corrected (see Figure 9). Kimura's "no-selection box" must be expanded significantly, firstly based on the imperfect correlation between genotype and phenotype, and then secondly based upon the imperfect correlation between phenotype and reproductive success. The result of these two imperfect correlations is that the number of near-neutral (un-selectable) nucleotide positions is greater than commonly realized, and their abundance is not strictly dependent on population size. Small population size certainly aggravates the noise problem, but large population size cannot eliminate this problem (for more detailed discussion see Appendix 5).

The pervasive existence of serious genetic noise amplifies virtually all my previous arguments regarding the limits of selection. Kimura's near neutral box gets bigger (Figure 9) - because huge numbers of otherwise minor mutations become near-neutral and hence un-selectable. The selection threshold problem, wherein simultaneous selection for too many traits results in a complete cessation of progress - will happen much sooner because of high levels of noise. Lastly, the noise problem will result in most "selection dollars" being completely wasted (Figures 8a-c). This greatly increases the actual cost of selection, and makes much more severe the very real limits that "cost" places upon any selection scenario.

To get a more intuitive understanding of the noise problem, we can return to some of our visual analogies. In terms of our evolving red wagon, try to imagine a situation where wagon performance was influenced much more by *workman errors* on the assembly line, than by the *typo-errors in the assembly manual.* Can you see that the quality control agent would mostly be wasting his time and resources - selecting for "non-heritable" variations? In terms of our Princess – the noise problem is like having innumerable pea-sized lumps within her mattresses. Wouldn't that make her problem worse? In terms of our biochemistry textbook, the noise problem is like having most of the typos in the textbook fail to be passed on into the next printing cycle. This would further reduce the already vanishingly small correlation between typographical errors and students' scores. In all these examples, when we add any reasonable level of noise, these already absurd scenarios just become even more impossible!

The late Stephen Jay Gould, like Kimura, argued against the strict selectionist view of evolution. In terms of the survival of entire species, he recognized the importance of natural disasters, "survival of the luckiest", and "noise". **What Gould and Kimura both seem to have failed to realize is that if noise routinely over-rides selection, then this makes long-term evolution impossible, and guarantees genetic degeneration and eventual extinction.**

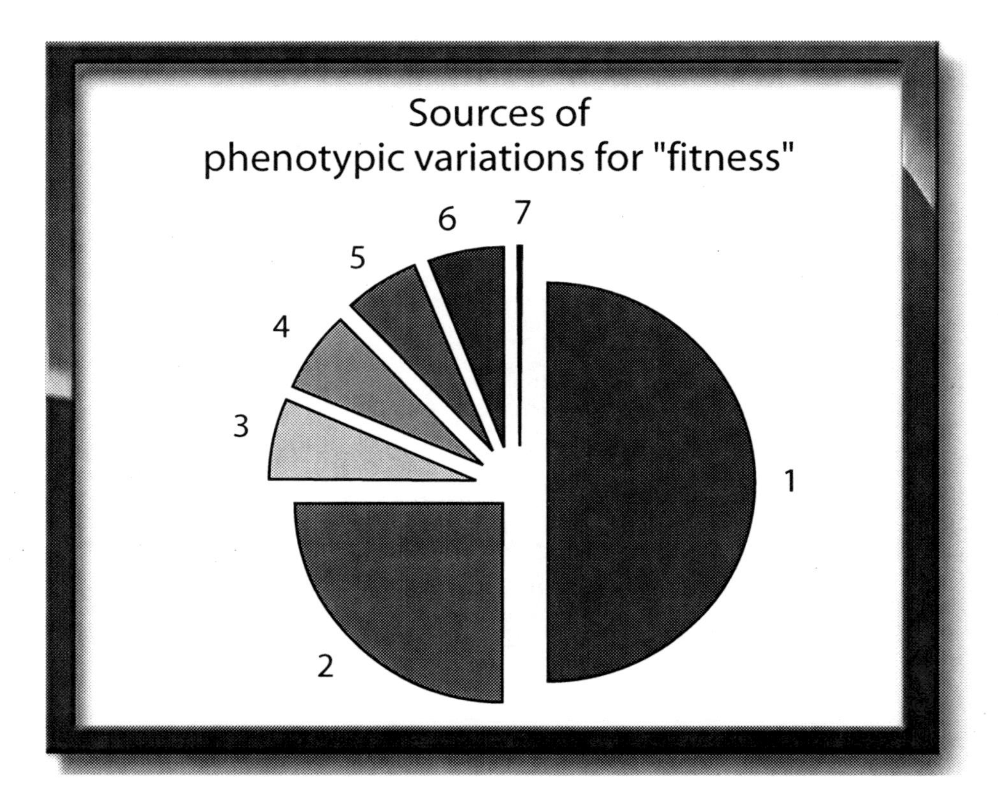

Figure 8a.

Sources of phenotypic variation. Variation between individuals results from numerous factors – both genetic and non-genetic. Within the genetic component of phenotypic variation, there are also numerous components. Only one genetic component is "selectable", and that is the "additive" genetic component. But this component is totally overshadowed by the other sources of phenotypic variation. The primary source of phenotypic variation is environmental variation (sector "1"). This variation is not heritable, and interferes with selection. The second major source of variation is the interaction of the environment with the genotype (2). This variation is also not heritable, and interferes with selection. Within the genetic component of variation, there is variation due to: epigenetics (3), epistasis (4), and dominance (5). None of these genetic components are heritable, and all of them interfere with true long-term selection. Lastly, there are "other" genetic components, which would otherwise be selectable, but are "neutralized", either by homeostatic processes or such things as cyclic selection (6). All these non-heritable components account for the vast bulk of all phenotypic variation. This leaves additive genetic variation as a relatively insignificant component of phenotypic variation (7). For a very general phenotypic trait, such as "reproductive fitness", additive variation can account for less than 1% of total phenotypic variation (Kimura, 1983, p.30-31). Another way of saying this is that more than 99% of any selective elimination based upon phenotypic superiority is *entirely wasted*. All variation that is not due to additive genetic variation, actually works very powerfully *against* effective selection! It acts as noise, which obscures the actual effects of heritable genetic variations.

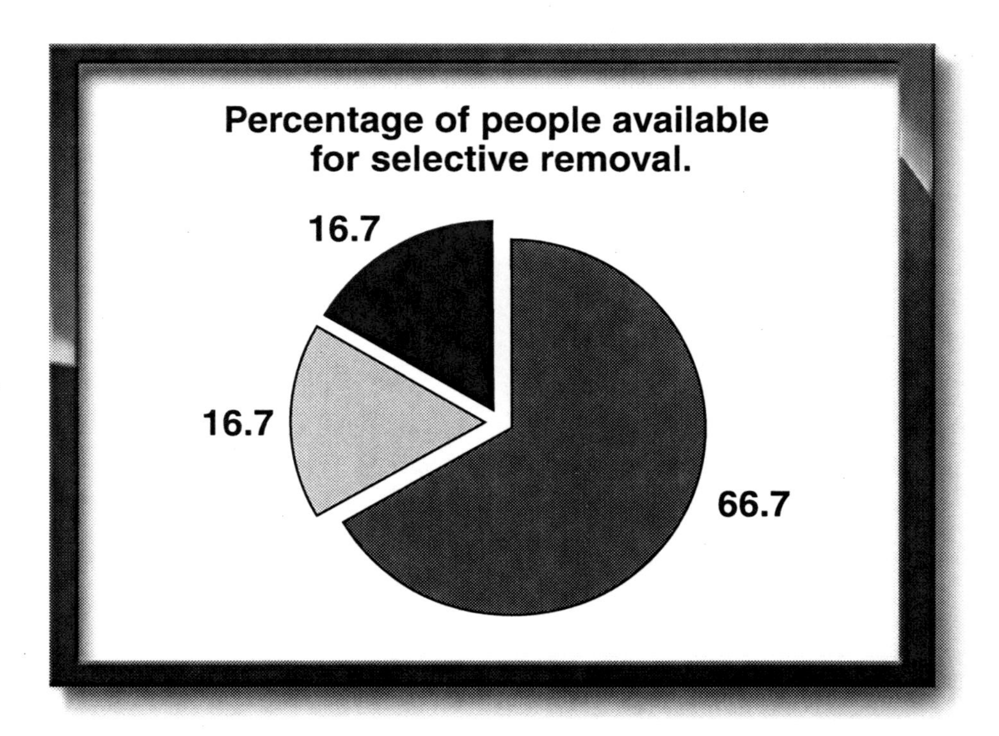

Figure 8b.

Percent population available for selective elimination. Only a limited proportion of any population (the *population surplus*) can be selectively eliminated. Given the current global rate of human reproduction (3 children per two adults), two-thirds of all children must reproduce for simple replacement. This maximally leaves one-third of the children available for selective reproductive elimination. However, a significant fraction (perhaps half) of the population's surplus will fail to reproduce for entirely random causes (i.e. war, accidents), which have nothing to do with "phenotype". So only about one-sixth (16.7%) of the human population is actually available for any potential selective elimination. This is in keeping with Haldane's estimate that only about 10% of a human population is actually available for selection.

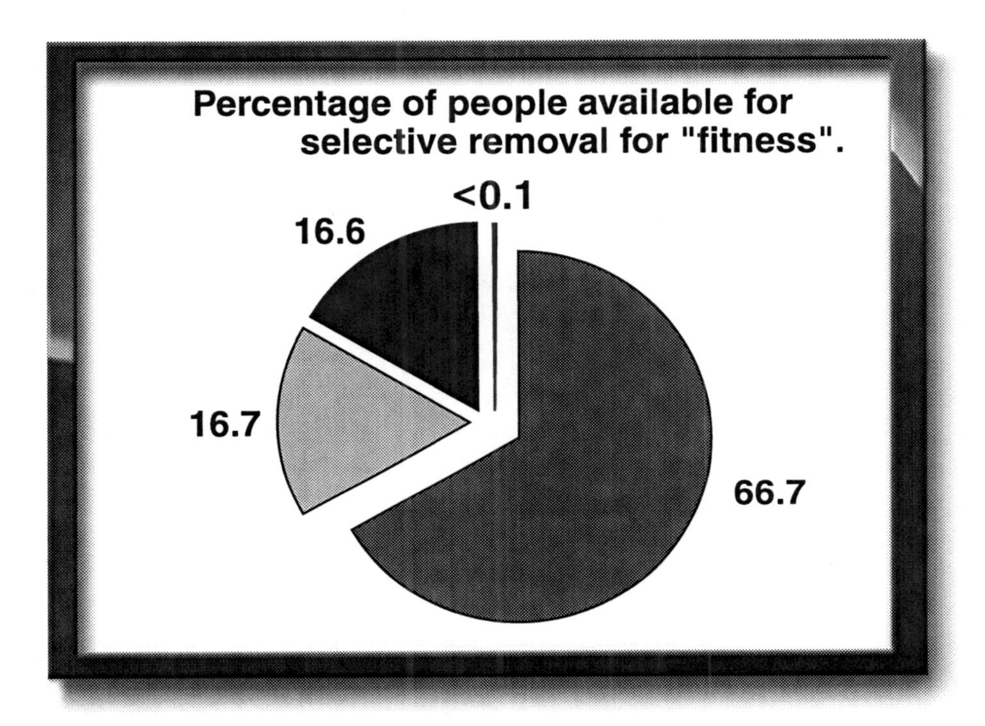

Figure 8c.

Selective elimination within human populations, based upon a very general trait such as "phenotypic fitness", has an effectiveness which approaches zero. This is seen when we combine Figures 8a and 8b. Of the 16.7% of a human population which is theoretically available for selective removal based upon phenotypic inferiority, only 0.4% of such removal will result in any heritable effect on succeeding generations. This means less than 0.1% (16.7% x 0.4% = 0.07%) of the total population is actually available for effective selection. This is too small a fraction to show accurately in this type of graph. In a sense, this means that less than one person in a thousand can be removed for truly *effective, selective, reproductive elimination.* So even if someone like Hitler were to "kill off" as many phenotypically "inferior" human beings as possible every generation, it would result in insignificant selective progress for something as general as "fitness". This conclusion is a logical extension of low human fertility rates and the extremely poor heritability of a trait as complex as fitness (Kimura, 1983, p.30-31).

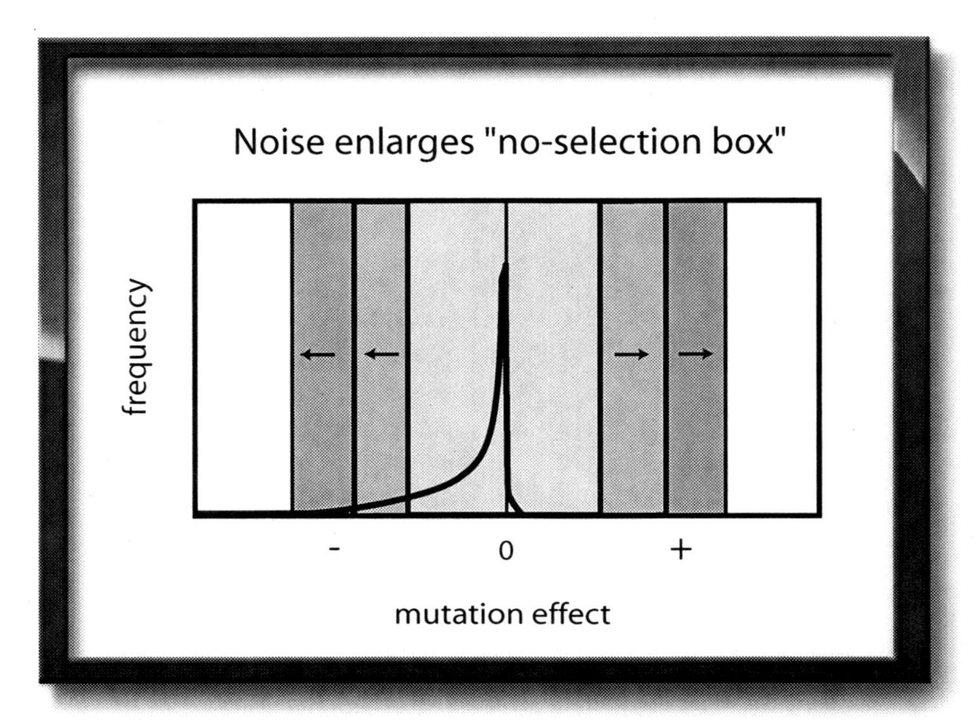

Figure 9

Kimura's near-neutral box, as shown in Figure 3d, is only based upon a minimal estimate of noise – i.e. only that attributable to gamete sampling. This is the classic near-neutral model. This view fails to recognize the primary sources of noise. So Kimura's classic near-neutral box is too small. We need to first expand our no-selection box because of poor heritability – which arises from the imperfect correlation between genotype and phenotype (more than 99% of phenotypic variation can be non-heritable (Figure 8a). We then need to expand our no-selection box further, in light of the imperfect correlation between phenotype superiority and reproductive success – which arises from the random aspects of reproduction (Figure 8b). These are the primary sources of noise. When we consider all sources of noise, we realize that the real "no-selection box" is large, and that it cannot be dismissed by simply invoking large population sizes.

Chapter 7

Crow to the rescue?

Newsflash – Crow solution fails reality test.

Is the net information within the genome going up or going down? We can, at best, wave our hands when we speculate about how selection might synthesize *new* information. It is inherently hypothetical. In a sense it becomes a philosophical question - not really subject to scientific analysis. Strong arguments can be made against mutation/selection creating new information, but theorists can always speculate to the contrary (it is very difficult to prove something can never happen). However, I believe the "going down" aspect of the genome is subject to actual scientific analysis. It is for this reason that I have focused on the issue of the degradation of information. I believe it is subject to concrete analysis. Such analysis persuasively argues that *net* information must be declining. If this is true, then even if it could be shown that there were specific cases where new information *might* be synthesized via mutation/selection, it would be meaningless - since such new information would promptly then begin to degenerate again. The net direction would still be *down*, and complex genomes could never have arisen spontaneously.

If the genome is actually degenerating, it is bad news for the long-term future of the human race. It is also bad news for evolutionary theory. If mutation/selection cannot *preserve* the information already within the genome, it is difficult to imagine how it could have *created* all that information in the first place! We cannot rationally speak of genome-building when there is a net loss of information every generation! Halting degeneration is just a small, prerequisite step, before the much more difficult question of *information-building* can reasonably be opened for discussion (see Chapter 9).

In the last decade, concern about the mutation rate has been growing among geneticists, as it has become more and more clear that the rate of deleterious mutations must be much higher than one per person per generation (Neel et al, 1986). One way for theorists to dismiss this problem has been to claim that most DNA is actually *non-information*, hence most mutations are perfectly neutral in effect. By this logic, if the actual rate of nucleotide substitution was 10 per person, then by defining 98% of the genome as junk DNA, the *effective* mutation rate must be only 0.2 per person. So the number commonly quoted in the media and textbooks would then be 0.2, not 10. However, as the known rate of mutation has been increasing, and as the recognized percent of functional DNA has been increasing, even this rationale has failed to explain away the problem of genetic degeneration. At this point, a distinguished geneticist, Dr. James Crow, described a model of natural selection which seemed to save the day (Crow, 1997).

Dr. Crow acknowledged that in any population, when the rate of deleterious mutations approaches 1 per individual, such mutations must begin to accumulate, and population fitness must decline. However, as the total number of accumulated mutations per person becomes quite large, he realized that some individuals would have significantly more mutations than others (due to chance). He proposed that by focusing selection against such individuals, one could sweep away a disproportionate number of mutations. The consequence would be that more mutations in the population would be eliminated at less "cost" to the population. Eventually, the number of mutants per person might then be stabilized, and the decline in fitness would taper off. This model seems to work in simple mathematical simulations, when it is assumed that all mutations are extremely minor, and all have an identical effect. An eventual leveling off of mutation accumulation is shown in a typical computer simulation - as summarized in Figure 10a. Assuming artificial "truncation selection", based solely upon "mutation count per individual", mutations accumulate to high numbers - but their increase eventually starts to taper off. However, the nature of this curve is surprising, in that it still shows a disastrous accumulation of mutations. It is very informative to look at the same simulation, as it affects fitness. In Figure 10b, we see what happens to fitness over time, even when we assign to each mutation the tiniest reasonable average value for mutation-effect (.0001) (if we lower this value much more, we would be in the near-neutral range, which would make all such mutants entirely un-selectable). What we see is that average fitness of the individuals in the population plummets essentially to zero (i.e. species extinction) in just 300 generations. What type of evolutionary scenario is this? This model has the appearance of extreme back bending, yet it still fails to stop catastrophic genomic degeneration. Moreover, the question remains – "does this highly artificial mathematical model match in any way what is really happening in nature?"

As mentioned earlier, we are all highly mutant, so selection must be between "more mutant versus less mutant" individuals. There are two ways of measuring who is "most mutant". The first, which is of minor importance, is the question - "Who has the most mutations?" The second, which is of primary importance, is - "Who has the worst mutations?" The model described by Crow only considers the former, while ignoring the latter. Hence the Crow model has very limited significance in the real world. It is an imaginary construct which seems to have been designed to obfuscate a major theoretical problem.

Because each of us is made up of genes that represent a more or less random sampling of the human "gene pool", we should each have very nearly the same total number of accumulated mutations per person. The actual difference in total number of mutations between two people will be quite trivial, only representing sampling variation - i.e. minor variations from the population's average. We know this because these mutations have been accumulating and mixing within the "gene pool" for many generations. Such mutants have grown very abundant, and mutation count should be quite homogeneous. Thus "mutation count" will be a very minor variable, in terms of differences between individuals.

If our genes are just samples from a large and well blended "gene pool" - why are we, as individuals, so distinct from one another? The answer to this question lies in the fact that different nucleotide differences have very different degrees of impact - ranging from lethal to neutral. True lethals reduce fitness 100% (zero reproduction). Typical minor mutations might reduce fitness by only 1 or 2%. Many 'near-neutrals' will only reduce fitness by a millionth of a percent or less. Because of all this, the fitness impact of different nucleotides can vary by *many orders of magnitude*. This means that the very striking genetic differences we see between human beings are not due to **total number** of mutants at all, but rather are due to the **specific effects** of relatively **few** *high-impact genetic differences.* Just one minor mutation can overshadow the effects of a *million* near-neutral mutations. One person may have several thousand fewer mutations than another, yet just one specific mutation can still make that person much less fit. Therefore, the idea of counting the total number of mutations per individual, and then selecting away high-count individuals can **not** be considered a reasonable natural mechanism to get rid of mutations. This concept appears to have been invented as a mathematical trick to attempt to rationalize how to get rid of more mutations, using less selection cost. Although this process may be operating to a very limited extent in nature, it is very clearly *not* what is generally happening.

Overwhelmingly, fitness variation between members of a population should be due to a relatively few, specific, high impact genetic variations - not total mutation count per person. A limited number of relatively major nucleotide variants are the real basis for the obvious differences we see between individuals and between races. Therefore, this limited number of "significant" nucleotides must be the real basis for all adaptive natural selection, and must also be the real basis for all artificial breeding of plants and animals. Even while selection can progress rapidly for a handful of variants at *major nucleotide sites*, the genome must still be degenerating *as a whole*, for all the reasons we have been discussing. Any appearance of genetic improvement is superficial (your car is still rusting away, even if you get a new coat of paint - the movie star is still aging, even if she gets a new facelift).

In dealing with the concepts above, one will encounter the term "synergistic epistasis". When I first encountered this phrase I was very impressed. In fact, I was intimidated. It seemed to speak of a very deep understanding, a deep knowledge, which I did not possess. As I have seen it used more, and have understood these issues better, I believe I understand the term better. It is a sophisticated-sounding expression, signifying nothing. It has all the appearance of deliberate *obfuscation*. Literally translated, synergistic epistasis means "interactive interaction". Does that help us? Fancy terminology is often used to hide lack of understanding. "Punctuated equilibrium" is an excellent example - it sounds so impressive, but explains so little. To the extent we can attribute any meaning to the term synergistic epistasis, it means that mutations interact such that several mutations cause more damage collectively than would be predicted by their individual effects. At least one paper provides experimental evidence that the concept is not valid (Elena and Lenski, 1997). But even if it were valid, it makes the genetic situation worse, not better. We have *always* known that genic units interact, and we know that such *epistasis* is a huge <u>impediment</u> to effective selection. This fact is normally conveniently ignored by most geneticists because selection scenarios become hopelessly complex and unworkable unless such interactions are <u>conveniently set aside</u>. But now, when genetic interactions can be used to obfuscate the problem of error catastrophe, the concept is conveniently trotted out and used in an extremely *diffuse and vague* manner - like a smoke screen. But let's look through the smoke. If multiple mutations really *do* create damage in a non-linear and escalating manner, then error catastrophe would happen much sooner and populations would spiral out of control much faster - into mutational meltdown. We would already be extinct! Once again we are looking at a conceptual "sleight of hand", which clearly does not apply to the real world, and which is only aimed at propping up the Primary Axiom.

Before leaving the subject of the Crow model, it should be pointed out that Crow's argument only addresses the problem of *cost of selection*. So even if Crow's model could be shown to be sufficient and fully operational in nature, the human genome should still deteriorate because of the many other reasons I have described - including the problems of near-neutral mutations, selection threshold for too many minor mutations, and selection interference.

By now we should clearly see that the Primary Axiom is not "inherently true", nor is it "obvious" to all reasonable parties - and so very clearly it should be rejected as an *axiom*. Moreover, what is left - the "Primary Hypothesis" (mutation/selection can create and maintain genomes) - is actually found to be *without any support*! In fact, *multiple lines of evidence indicate that the "Primary Hypothesis" is clearly false*, and must be rejected.

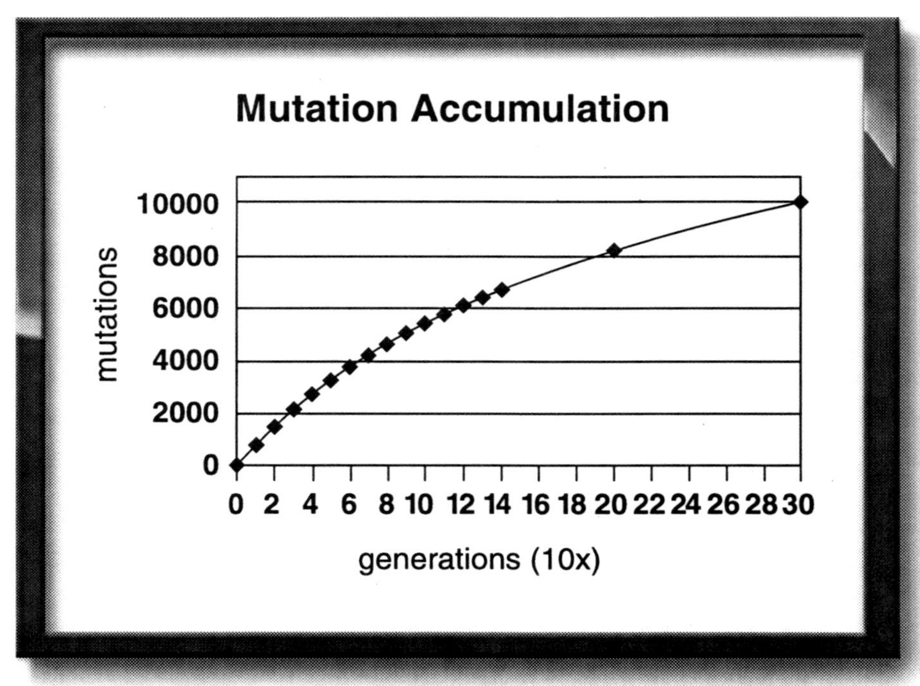

Figure 10a.

Crow's mutations. At my request, Walter ReMine has kindly developed software for doing numerical simulations of Dr. Crow's model of "truncation selection based on mutation count". The curve shown above plots average number of mutations accumulated per person, after (n) generations – assuming sexual recombination, 100 individuals in the population, 100 mutations per person, 4 offspring per female, 25% non-genetic (random) elimination, and 50% selective elimination of the remainder, per generation. Although the rate of mutation accumulation eventually begins to level off, this does not happen until very serious genetic damage has been done. Given a more realistic model, there is no reason to expect the mutation count to level off. Crow's model is designed to make the problem of mutation accumulation "go away". It assumes all mutations have equal value, are all individually very subtle, yet none are so subtle as to be "nearly neutral", that all selection is based upon "mutation count", and that artificial "truncation selection" is operational. None of these assumptions are even remotely reasonable. Even though all these assumptions are artificial and unreasonable, the numerical simulation still shows severe mutation accumulation. Almost identical mutation accumulation curves have been modeled by Schoen et al., 1998.

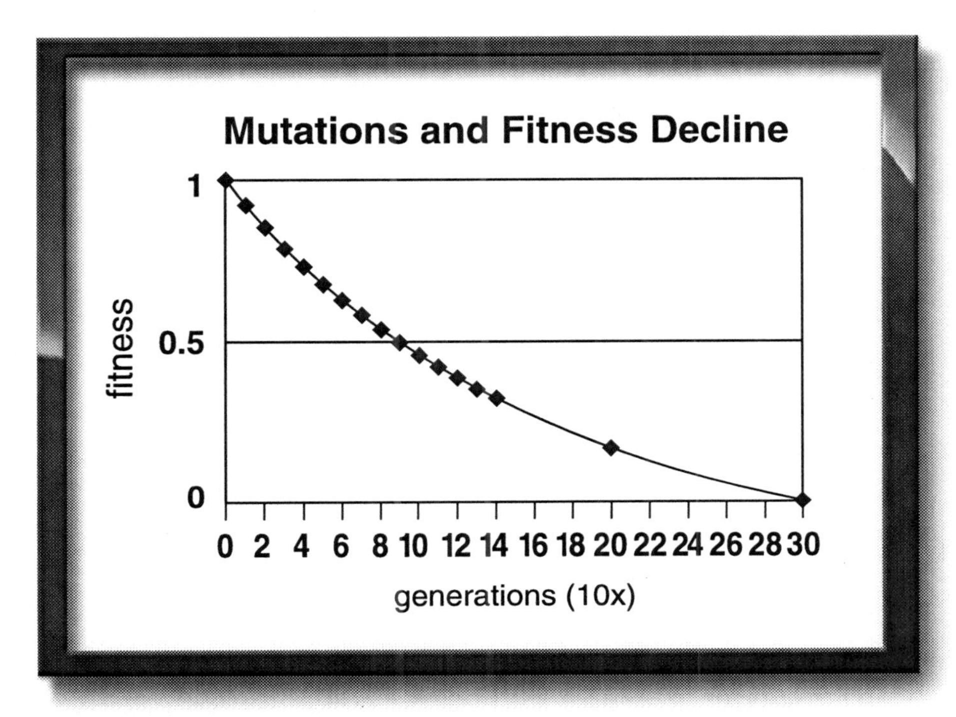

Figure 10b.
Crow's fitness decline. Using the data shown in 10a, we can plot average population fitness over time, assuming Crow's model of "truncation selection based on mutation count". We must assign an average value to the mutations that are accumulating. We can assume the average mutation value is at least .0001 (each mutation reduces fitness only one part in ten thousand). This level of "average mutation-effect" is very conservative. As seen above, as mutations accumulate, the average fitness naturally declines. Assuming an additive model, the result is that our species goes extinct in roughly 300 generations. Yet if we reduce the average mutation-effect to substantially less than .0001, we would arguably be making all the mutations effectively neutral, and therefore un-selectable. If the "average mutation" actually becomes effectively neutral and un-selectable, then Crow's model breaks down completely, and there can be no effective selection strategy to stop mutation accumulation. Schoen et al. (1998) have modeled almost identical fitness decline curves, which arise from mutation accumulation.

Man to the rescue?

Newsflash – Eugenics cannot stop degeneration, and cloning makes the mutation problem worse.

Most informed geneticists would acknowledge that we currently have a problem, in terms of genetic degeneration of the human genome, due to relaxed selection. They would probably even acknowledge that this problem extends to many other species – certainly it applies to all endangered species. They would probably even acknowledge the theoretical difficulties of establishing an effective selection scheme that would stop the accumulation of genetic damage within the human population. However, many would also point out that the genetic degeneration process is gradual – with only a 1-2% decline in fitness per generation. So they might say that the problem is not urgent. While it is true that the ultimate consequences of today's mutations will not be fully felt for many generations, it is probably equally true that in order to halt the overall genetic degeneration, the time to stop it, if it could be stopped, would be now. As the problem becomes more advanced, genetic damage should tend to get way ahead of us, with a potential run-away (meltdown) situation developing. For those hedonists who live only for today – so what? But for those idealists who pin all their hopes on the continual survival and advancement of the human race – the issue should be of great importance. They are likely to turn hopeful eyes to science and technology to solve the problem. Can human intervention triumph over the present threat? The nature of the threat of genetic degeneration is such that the accumulating damage is inherently diffused throughout the genome – it cannot be dealt with one gene at a time. It should therefore be obvious that the laser-like precision of genetic engineering has nothing to offer this situation. However, we might reasonably ask if artificial selection, or the emerging capacity to clone human beings, might not solve the problem.

Eugenics to the rescue?

The general perception that man is degenerating is found throughout modern and ancient literature. All cultures have legends about "men of old", who were smarter, more powerful, and longer-lived. Darwin's book, *Origin of Species and the Survival of Favored Races*, introduced the new idea that strong and continuous selection ("survival of the fittest") might halt this perceived degenerative trend. Darwin repeatedly pointed to human efforts in animal and plant breeding as a model for such man-directed selection. In his book *The Descent of Man*, Darwin went further, and contended that there is a need for superior races (i.e. the white race) to replace the "inferior races". This ushered in the modern era of racism, which came to a head in Hitler's Germany. Before World War II, many nations, including America, had government-directed eugenics programs. These governmental programs included forced sterilization of the "unfit", and aggressive promotion of abortion/fertility-control for the underclasses. Ever since the time of Darwin, essentially all of his followers have been eugenicists at heart, and have advocated the genetic improvement of the human race. When I was an evolutionist, I also was, at heart, a eugenicist. The philosophers and scientists who created the modern "synthetic theory" of evolution were uniformly eugenicists. However, after the horrors of WWII, essentially all open discussions of eugenics were quietly put aside.

In the new light of a deteriorating human genome, should eugenics be re-examined? Unfortunately, this is already happening. However, this is neither morally nor scientifically defensible. The thesis of this book cannot logically be used to support such efforts, but rather argues against such efforts. The eugenicist's vision is a delusion, and one which is insidious. The central thesis of this book is that no form of selection can stop genomic degeneration. This includes artificial selection. To attempt to artificially increase selection intensity globally would be physically, socially, and politically impossible. Only a ruthless world-wide authoritarian "power elite" could possibly dictate the standards for selection - deciding who could and who could not reproduce. Even if such an insidious plan could effectively be implemented, it would not significantly slow genomic degeneration. Any potential genetic "progress" would be trivial, and would not be sufficient to offset the overall degeneration of the genome. "Inferiority" versus "superiority" is an ambiguous and very poorly inherited characteristic, and is largely influenced by non-genetic (environmental) factors (see Chapter 6). Selection for any non-genetic trait results in zero selective progress. Even selection for more "heritable" traits would be ineffective, because of all the genetic arguments I have just presented. "Fitness" is a very general term, and is affected by countless genes, most of which are minor or near-neutral. As we have been learning, effective selection can only act upon a handful of "significant" genes. It is true that we could artificially select for virtually any single human trait - to make people taller or shorter, lighter or darker, or fatter or skinnier. But we could not effectively select for *superior* - which inherently involves thousands of genes and millions of nucleotides.

Any possible *benefit* of enhanced or artificial selection in man would be a slight improvement of a very limited number of very specific traits. But the genome would still be "rusting out" - and at about the same overall speed. The *cost* of eugenics would be social and moral upheaval on a level that would be catastrophic. Eugenics has from its inception been a racist concept, and has always been driven by the Primary Axiom. Eugenics is not genetically sound. Furthermore, it is tightly linked with authoritarian government, elitist philosophy, suppression of personal rights, and violation of human dignity. Eugenics cannot rescue us from genomic degeneration. If we foolishly try to rescue the genome in this way - then who will then rescue us from the eugenicists?

Cloning to the rescue?

To clone a human being means to take a cell from a mature (presumably superior) individual, and to use that cell to produce one or more copies of that individual. The consequence of cloning would have profound genetic consequences. While I was still an evolutionist and was very concerned about the genetic degeneration of man, I naively believed that cloning might be the answer. I hoped that cloning might halt genomic degeneration, and might even allow rapid improvement of the human population.

In plant genetic improvement, when a species is easily propagated clonally, clonal selection provides the surest and fastest way to improve a population. Just choose the best individuals, and multiply! It is about that simple. Given this knowledge, perhaps it is surprising that clonal selection for man has not been more vigorously advocated. However, even apart from the moral and social problems associated with cloning, the best-case scenario for cloning would involve possible short-term genetic gains, but would guarantee long-term genetic degeneration. A central theme of this book, if restated, would be that what is already known for clonal populations also applies to sexual populations. The proof that all clonal populations must degenerate genetically was first shown by Muller, and has been termed "Muller's ratchet" (Muller, 1964; Felsenstein, 1974). Basically, within any clonal line, even the best sub-clones will accumulate mutations over time, and there is no mechanism to "clean house". Even selection within a clonal line for the *best sub-clones*, does not stop the decline. The "ratchet" only works one way - all change must be downward. Each cell division adds mutations, and there is no mechanism to take mutations away. Even allowing for some beneficial mutations, each beneficial mutant is always linked, within its sub-clone, to many, many more deleterious mutations. Within clones, there is no mechanism to break this linkage between rare beneficial mutations and abundant deleterious mutations. The deleterious mutants will always grow in number much faster than the beneficials, and so they will always drag any beneficials down with them. Therefore net information must always decline. To repeat - this applies to even the best, most highly selected sub-clones. The certainty of genomic degeneration in clonal populations is well known among geneticists. The reason there are still many populations of clonal plants (and some animals), appears to reflect the fact that there has not yet been enough time for such populations to degenerate to the point of extinction.

Preliminary animal cloning experiments indicate the cloning of animals cannot even produce short-term genetic gains. Cloned animals routinely display immediate and severe genetic damage. Why is this? Cloned animals routinely show evidence of mutational damage - as if they are "pre-aged" (Hochedlinger et al., 2004). There are probably multiple reasons for this (genetic and epigenetic), but one major reason involves the fact that mutations continue to build up within somatic cells. Normal reproductive cells are "designed" to undergo the least possible cell divisions between generations, thereby minimizing mutation accumulation per generation. It is for such reproductive cells, that we have the minimal estimate of "just" 100 point mutations per person per generation. For somatic cells (at least for the continuously-dividing stem cells), the number of cell divisions is much higher than in the germline, and so the mutation rate should also be very much higher - sharply increasing with age. As we grow and then begin to age, every cell in our body becomes genetically more mutant and more "unique". It is impossible to keep a large genome unchanged, even within a clone! New mutations occur in every cell, at the rate of roughly *one mutation every cell division*. Therefore, essentially every single cell in our body is unique. For these reasons every human clone will always be inferior to the mature "source individual" from which they were cloned. Such a clone will in a sense be *pre-aged* - having the original mutational load of the source individual, plus the mutational load that has accumulated during that person's growth and aging. Because of the many cell divisions during somatic development and stem cell maintenance, a human clone will be roughly comparable, in terms of degeneration, to individuals many sexual generations into the future. In this sense such a clone is like a foreshadowing of where the species is going. It is going down, not up (see Figure 11).

There are powerful moral, social, and genetic arguments against cloning. Cloning can be viewed as a very high-tech form of eugenics - with all its technical and moral problems. Eugenics in general, and cloning in particular, are definitely <u>not</u> solutions to genomic degeneration.

Figure 11.

Degeneration of the genome, degeneration of man, and degeneration of mankind. We experience it on a personal level, and we see it all around us. It is "genetic entropy", and there is nothing man can do to halt it. It is biologically inevitable. It is why species go extinct, and it is why we are all individually in the process of dying.

Chapter 9

Can natural selection create?

Newsflash - Mutation/selection cannot even create a single gene.

We have been analyzing the problem of genomic degeneration and we have found that regardless of how we analyze it, the genome must clearly degenerate. This problem overrides all hope for the forward evolution of the whole genome. However some limited traits might still be improved via mutation/selection. Just how limited is such progressive ("creative") mutation/selection? From the perspective of our analogy, an instruction manual, we can intuitively see that not even a single component of a jet plane (let's say a molded aluminum component) could realistically arise by misspellings within the manual. So it is certainly reasonable to then ask the parallel question - "Could mutation/selection create even a single functional gene?" The answer is that it cannot - because of the enormous preponderance of deleterious mutations, even within the context of a single gene. The net information must always still be declining, even within a single gene. However, to better understand the limits of forward selection, let us for the moment *discount all deleterious mutations* and only consider beneficial mutations. Could mutation/selection *then* create a new and functional gene?

1. Defining our first desirable mutation - The first problem we encounter in trying to create a new gene via mutation/selection is defining our first beneficial mutation. By itself, no particular nucleotide (A, T, C, or G) has more value than any other - just as no letter in the alphabet has any particular meaning outside of the context of other letters. So selection for any single nucleotide can never occur, except in the context of all the surrounding nucleotides (and, in fact, within the context of the whole genome). Like changing a letter within a word or chapter, the change can only be evaluated in the context of all the surrounding letters. We cannot define any nucleotide as good or bad except in relation to its neighbors and their shared functionality. This brings us to an excellent example of the principle of "irreducible complexity". In fact, it is irreducible complexity at its most fundamental level. We immediately find we have a paradox. To create a new function, we will need to select for our first beneficial mutation, but we can only define that new nucleotide's value in relation to its neighbors. Yet to create any new function, we are going to have to be changing most of those neighbors also! We create a circular path for ourselves - we will keep destroying the "context" we are trying to build upon. This problem of the fundamental inter-relationship of nucleotides is called *epistasis*. True epistasis is essentially *infinitely complex*, and virtually impossible to analyze, which is why geneticists have always conveniently ignored it. Such bewildering complexity is exactly why language (including genetic language) can never be the product of chance, but requires intelligent design. The genome is literally a book, written literally in a language, and short sequences are literally sentences. Having random letters fall into place to make a single meaningful sentence, by accident, is numerically not feasible. The same is true for any functional strings of nucleotides. If there are more than several dozen nucleotides in a functional sequence, we know that realistically they will *never* just "fall into place". This has been mathematically demonstrated repeatedly. But as we will soon see, neither can such a sequence arise randomly one nucleotide at a time. A pre-existing "concept" is required as a framework upon which a sentence or a functional sequence must be built. Such a concept can only pre-exist within the "mind of the author". Starting from the very first mutation, we have a fundamental problem - even in trying to define what our first desired beneficial mutation should be!

2. Waiting for the first mutation - Human evolution is generally assumed to have occurred in a small population of about 10,000. The mutation rate for any given nucleotide, per person per generation is exceedingly small (only about one chance in 30 million). So in a typical evolutionary population, if we assume 100 mutations per person per generation, one would have to wait 3,000 generations (at least 60,000 years) to expect a specific nucleotide to mutate - within a population of 10,000. But two out of three times, it will mutate into the "wrong" nucleotide. So to get a specific desired mutation at a specific site will take three times as long - or at least 120,000 years. Once the mutation has occurred, it has to become *fixed* (such that all individuals in the population will have two copies of it). For new mutations, because they are so rare within the population, they have an extremely great probability of being *lost* from the population, due to random genetic drift. Only if the mutation is dominant and has a very distinct benefit does selection have any reasonable chance to rescue any given new mutation from random elimination via drift. According to population geneticists, apart from effective selection, in a population of 10,000 our given new mutant has only one chance in 20,000 (the total number of non-mutant nucleotides present in the population) of **not** being lost via drift. Even with some modest level of selection operating, there is a very high probability of random loss, especially if the mutant is recessive or is weakly expressed (we actually know that almost all beneficial mutations will be both recessive and nearly-neutral). For example, if a mutation increases fitness by half of one percent, it only has a 1% probability of becoming fixed. So realistically, at least 99 out of 100 times the desired beneficial mutation will be randomly lost. So a typical mildly-beneficial mutation must happen about 100 times before it is likely to "catch hold" within the population (even though it is beneficial!). So on average, we would have to wait 120,000 x 100 = 12 million years to stabilize our typical first desired beneficial mutation, to begin building our hypothetical new gene. So, in the time since we supposedly evolved from chimp-like creatures (6 million years), there would not be enough time to realistically expect our first desired mutation – the one destined for fixation.

3. Waiting for the other mutations - After our first mutation has been found (the one that will eventually be fixed), we need to repeat this process for all the other nucleotides encoding our hoped-for gene. A gene is minimally 1,000 nucleotides long (this is really 50-fold too generous - I am ignoring all regulatory elements and introns). So if this process was a straight, linear, and sequential process - it would take about 12 million years x 1,000 = 12 billion years to create the smallest possible gene. This is approximately the time since the reputed big bang! **So it is a gross understatement to say that the rarity of desired mutations limits the rate of evolution!**

4. Waiting for recombination - Because sexual species (such as man) can shuffle mutations, it might be thought that all the needed mutations for a new gene might be able to occur simultaneously within different individuals within the population, and then all the desirable mutations could be "spliced together" via recombination. This would mean that the mutations would not have to occur sequentially - shortening the time to create the hoped-for gene (so we might need less than billions of years). There are two problems with this. Firstly, when we examine the human genome, we consistently find the genome exists in large blocks (20,000-40,000 nucleotides) wherein no recombination has occurred - *since the origin of man* (Gabriel et al. 2002, Tishkoff and Verrelli, 2003*)*. This means that virtually no meaningful shuffling is occurring on the level of individual nucleotides. Only large gene-sized blocks of DNA are being shuffled. **I repeat - no actual *nucleotide* shuffling is happening!** Secondly, even if there were effective nucleotide shuffling, the probability of getting all the mutants within the population to shuffle together into our hoped-for sequence of 1,000 is so astronomically remote that we would need even *more* time than by the sequential approach (even *more* billions of years) for this scenario to work. Lastly, if there really *were* this type of extensive "nucleotide shuffling", which might build a new gene in this way, the very first generation after the new gene fell into place, it would be torn apart again by the same extensive nucleotide shuffling. In poker, it is not likely you will be dealt a royal flush. If you are, and then the cards are re-shuffled - what are the odds you will then get that very same hand dealt to you again?

5. Waiting on "Haldane's dilemma" - Once that first mutation that is destined to become fixed within the population has finally occurred, it needs time to undergo selective amplification. A brand new mutation within a population of 10,000 people, exists as only one nucleotide out of 20,000 alternatives (there are 20,000 nucleotides at that site, within the whole population). The mutant nucleotide must "grow" gradually within the population, either due to drift or due to natural selection. Soon there might be two copies of the mutant, then 4, then 100, and eventually - 20,000. How long does this process take? For dominant mutations, assuming very strong unidirectional selection, the mutant might conceivably "grow' within the population at a rate of 10% per generation. At this very high rate, it would still take roughly 105 generations (2,100 years) to increase from 1 to 20,000 copies ($1.1^{105} = 20{,}000$). However, in reality mutation fixation takes very much longer than this, because selection is generally very weak, and most mutations are recessive and very subtle. When the mutation is recessive, or when selection is not consistently unidirectional or strong, this calculation is much more complex – but it is obvious that the fixation process would be very dramatically slower. For example, an entirely recessive beneficial mutation, even if it could increase fitness by as much as 1%, would require at least 100,000 generations to fix (Patterson, 1999).

A famous geneticist, Haldane (1957), calculated that given what he considered a "reasonable" mixture of recessive and dominant mutations, it would take (on average) 300 generations (at least 6,000 years) to select a single new mutation to fixation. Selection at this rate is so very slow, it is essentially the same as no selection at all. This problem has classically been called "Haldane's dilemma". At this rate of selection, one could only fix 1,000 beneficial nucleotide mutations within the whole genome, in the time since we supposedly evolved from chimps (6 million years). This simple fact has been confirmed independently by Crow and Kimura (1970), and ReMine (1993, 2005). The nature of selection is such that selecting for one nucleotide always reduces our ability to select for other nucleotides (selection interference) - therefore *simultaneous selection does not hasten this process.*

At first glance, the above calculation seems to suggest that one *might* at least be able to select for the creation of one *small* gene (of up to 1,000 nucleotides) in the time since we reputedly diverged from chimpanzee. There are two reasons why this is not true. First, Haldane's calculations were only for independent, unlinked mutations. Selection for 1,000 *specific and adjacent* mutations could not happen in 6 million years - because that specific sequence of adjacent mutations would never arise - not even in 6 *billion* years. One cannot select mutations that have not happened. Secondly, as we will soon see, the vast bulk of a gene's nucleotides are near-neutral and cannot be selected at all – not in *any length of time.* The bottom line of Haldane's dilemma is that selection to fix new beneficial mutations occurs at glacial speeds, and the more nucleotides which are under selection, the slower the progress. This severely limits progressive selection. Within reasonable evolutionary timeframes, we can only select for an extremely limited number of unlinked nucleotides. In the last 6 million years, selection could maximally fix 1,000 unlinked beneficial mutations – creating less new information than is on this page of text.* There is no way that such a small amount of information could transform an ape into a human.

Although we have temporarily suspended deleterious mutations from consideration, it is only fair now to note that within the same timeframe that we hypothetically evolved from chimps, geneticists

** In terms of information content, 3 nucleotides equal roughly 1 typewritten character (there are only 4 nucleotides, but 26 letters, and more than 64 keys on a keyboard). So one codon triplet equals roughly one typographical "letter", and thus 1000 nucleotides equals only 333 spaces on a typewritten page.*

believe that many thousands of deleterious mutations should have been also fixed, via genetic drift (Kondrashov, 1995; Crow, 1997; Eyre-Walker and Keightley, 1999; Higgins and Lynch, 2001). Therefore, our evolutionary assumptions should lead us to logically conclude that we should have significantly degenerated downward from our ape-like ancestors (deleterious fixations greatly outnumbering beneficial fixations). The power of this logic is overwhelming. In fact, we know man and chimp differ at roughly 150 million nucleotide positions (Britten, 2002), which are attributed to at least 40 million hypothetical mutations. Therefore, assuming man evolved from a chimp-like creature - during that process there must have been about 20 million nucleotide fixations within the human lineage (40 million divided by 2), yet we now can see that natural selection could only have selected for 1,000 of these mutations. All the rest (about 20 million) would have had to have been fixed by random drift - resulting in millions of nearly-neutral deleterious substitutions. The result? A maximum of 1000 beneficial substitutions - in opposition to millions of deleterious substitutions. This would not just make us inferior to our chimp-like ancestors, it would obviously have killed us!

6. Endless fitness valleys - Evolutionists agree that to create a new gene requires a great deal of mutational "experimentation". During the "construction phase" of developing a new trait or a new gene, we have to expect a period of time when the experiment reduces a species' fitness. This is called a *fitness valley*. A half-completed gene is neither beneficial nor neutral - it is going to be deleterious. So in a sense, the species has to get worse before it can get better. It is easy to imagine a species surviving fitness valleys if they are brief, and if they are rare.

However, long deep fitness valleys are likely to lead to extinction, not evolution. The extreme rarity of desired mutations and the extreme slowness resulting from Haldane's dilemma, should make fitness valleys indefinitely long and deep. If evolutionary innovation was continuous (as is widely claimed), then a species' fitness should just keep going down - life would be just one fitness valley upon another upon another. Life's super-highway of evolution would always be under construction, and total fitness would always be declining rather than increasing. The concept of a species passing through fitness valleys makes evolutionary sense only when individual traits are considered. However, when the whole genome is considered, the concept of indefinitely numerous, and indefinitely long, fitness valleys argues strongly against the evolution scenario.

7. Poly-constrained DNA - Most DNA sequences are *poly-functional*, and so must also be *poly-constrained*. This means that when a DNA sequence has meaning on several different levels (poly-functional), each level of meaning limits possible future change (poly-constrained). For example, imagine a sentence which has a very specific message in its normal form, but has an equally coherent message when read backwards. Now let's suppose that it also has a third message when reading every other letter, and a fourth message when a simple encryption program is used to translate it. Such a message would be poly-functional and poly-constrained. We know that misspellings in a normal sentence will not normally improve the message - but at least this would be *possible*. However, a poly-constrained message is fascinating, in that it cannot be improved - it can *only* degenerate (see Figure 12). Any misspellings which might possibly improve the normal sentence form - will be disruptive to the other levels of information.

Any change at all will diminish total information - with absolute certainty.

There is abundant evidence that most DNA sequences are poly-functional, and therefore are poly-constrained. This fact has been extensively demonstrated by Trifonov (1989). For example, most human coding sequences encode for two different RNAs, read in opposite directions (i.e. both DNA strands are transcribed – Yelin et al., 2003). Some sequences encode for different proteins depending on where translation is initiated and where the reading frame begins (i.e. read-through proteins). Some sequences encode for different proteins based upon alternate mRNA splicing. Some sequences serve simultaneously for protein-encoding and also serve as internal transcriptional promoters. Some sequences encode for both a protein coding region, and a protein-binding region. Alu elements and origins-of-replication can be found within functional promoters and within exons. Basically all DNA sequences are constrained by isochore requirements (regional GC content), "word" content (species-specific profiles of di-, tri-, and tetra-nucleotide frequencies), and nucleosome binding sites (i.e. all DNA must condense). Selective condensation is clearly implicated in gene regulation, and selective nucleosome binding is controlled by specific DNA sequence patterns - which must permeate the entire genome. Lastly, probably all sequences do what they do, even as they also affect general spacing and DNA-folding/architecture - which is clearly sequence dependent. To explain the incredible amount of information which must somehow be packed into the genome (given that extreme complexity of life), we really have to assume that there are even higher levels of organization and information encrypted within the genome. For example, we know there is another whole level of organization at the epigenetic level (Gibbs, 2003). There also appears to be extensive sequence-dependent three-dimensional organization within chromosomes and the whole nucleus (Manuelidis, 1990; Gardiner, 1995; Flam, 1994). Trifonov (1989), has shown that probably all DNA sequences in the genome encrypt multiple "codes" (up to 12 codes). In computer science, this type of "data compression" can only result from the highest level of information design, and results in maximal information density. These higher levels of genomic organization/information content, greatly multiply the problem of poly-constrained DNA. Every nucleotide interacts with many other nucleotides, and everything in the genome seems to be context-dependent. The problem of ubiquitous, genome-wide, poly-constrained DNA seems absolutely overwhelming for evolutionary theory. Changing *anything* seems to potentially change *everything*! The poly-constrained nature of DNA serves as strong evidence that higher genomes cannot evolve via mutation/selection - except on a trivial level. Logically, all poly-constrained DNA had to be designed.

8. Irreducible complexity - The problem of irreducible complexity has been brilliantly presented by Behe (1996). He has illustrated the concept of irreducible complexity in various systems that have multiple components, such as a mousetrap design which requires 5 independent parts, or a flagellum having perhaps 10-20 component parts. His idea is that each *part* has no value except within the context of the *whole* functional unit, and so irreducible systems have to come together all at once, and cannot arise one piece at a time. In the case of a mousetrap - all the pieces may have been sitting next to each other on the inventor's workbench - but they would not have come together by chance, or by any realistic evolutionary progression. They came together as a *synthesis*, simultaneously, in the mind of the inventor. It is in the realm of *mind* that deep complexity comes together and becomes integrated.

In our example of the evolution of transportation technology, the simplest first improvement we might imagine might be the occurrence of misspellings that would convert our red wagon into a blue tricycle. It is indeed easy to imagine a misspelling that might cause the paint code to be changed (although the blue paint would have to already be available, and coded). Likewise, a misspelling could certainly cause a wheel to fall off. However, a three-wheeled wagon is not a tricycle – it is a broken wagon. To convert a wagon to a trike would require extensive reworking of the instruction manual and radical changes in most of the manufactured component parts. There would be no intermediate functional steps to accomplish these complex changes, and so no prospect for our quality control agent to selectively help the process along – in fact he would be selecting against all our desired misspellings and changes. So the correct combination of misspellings would have to arise simultaneously by chance, all at the same time – which would never ever happen. Obviously, a trike could only arise from a wagon by way of intelligent and extensive reworking of the design, and a thorough re-writing of the instruction manual (see Figure 13).

Although a wagon or trike may have dozens of component parts, even the simplest protein is a much more complex machine - having hundreds of component parts, and thus representing irreducible complexity profoundly greater than that illustrated by our wagon analogy. As the number of components of a design increases linearly, the number of *interactions* (hence the complexity) increases exponentially.

As complex as proteins are, underlying every protein is a genetic system comprising even higher levels of irreproducible complexity.

The molecular machinery underlying the coding, transcription, and translation of a protein is phenomenal. Ignoring all the other accessory proteins involved, just the design of the DNA/RNA sequence is mind-boggling. Although a simple protein has a few hundred component parts, the underlying gene that produces it has thousands of component parts. All of these parts are interacting and mutually-defining. Each nucleotide has meaning only in the context of all the others. Each nucleotide is poly-functional - interacting with many other nucleotides. The DNA sequence defines regional 3-D chromatin structure, local protein binding, uncoiling, transcription, and also defines one or more RNA sequences. The RNA sequence defines RNA stability, RNA variable splicing, RNA processing, RNA transport, transcription efficiency, and protein sequence.

We do not yet really understand how any single gene from a higher life form really works - not in its entirety. Not in the context of everything else that is happening in the cell. A single gene with all its interactions is still way too complex for us. When we consider the full complexity of a gene, including its regulatory and architectural elements, a single gene has about 50,000 component parts. I presume that this is more component parts than are found in a modern automobile. There is no simple linear path that leads car components to spontaneously become a functional car - *mind* is obviously required (actually, *many* brilliant minds). In the same way, there is no linear path of selection that can build a single gene from its individual nucleotides – a *mind* is likewise required. **Yet a single gene is just a microscopic speck of irreducible complexity, within the universe of irreducible complexity that comprises a single cell. Life is itself the very essence of irreducible complexity - which is why we cannot even begin**

to think of creating life ourselves. Life is layer upon layer upon layer of irreducible complexity. Our best biochemical flow charts, of which we are so proud, are just childish cartoons of true biological complexity - which is something we cannot even comprehend. It is a tribute to the mind of man that we have started to understand how even a single gene works, and that we can now design and build very small artificial genes. But we still cannot design a new gene for a new and unknown protein, which could then precisely integrate into the complexity of a higher life form. If we cannot do this - why would we think that random mutations, combined with a very limited amount of reproductive sieving, could accomplish this? For the reader's interest I have attempted to expand upon the concept of irreducible complexity - with the concept of Integrated Complexity (see Appendix 3).

9. Almost all beneficial mutations must be near-neutral. We have already discussed at length the difficulty of selecting against near-neutral deleterious mutations, and this problem is begrudgingly acknowledged by most geneticists. However, there is a flip side to this problem, which is even more important, but which I have *never* heard acknowledged. As we have already discussed in Figure 3d, the problem of near-neutrality is much more severe for beneficial mutations than for deleterious mutations. Essentially *every* beneficial mutation must fall within Kimura's "no selection zone". All such mutations can *never* be selected for. This problem multiplies all of the problems I have already outlined above. Our hoped-for new gene will certainly have a *few* nucleotides that have major effects - for example the ones that specify the active site of an enzyme. But such nucleotides can only have major effects within the context of the whole protein and the whole gene sequence. The whole protein/gene is constructed primarily with components that individually have only a small impact on the whole unit, and have only a miniscule impact on the fitness of the whole individual. In combination, these nucleotides contain most of the information contained within the gene - without them the "important nucleotides" are meaningless. Yet they are all individually un-selectable. So how can we establish them and keep them in their respective places, during gene construction? The answer is obviously that we simply cannot. And apart from these "insignificant masses" of nucleotides the elite "important nucleotides" cannot be selected for either. Because of the near-neutral problem, we cannot even get to first base in terms of building our hoped-for new gene. The entire framework of the new gene is defined by the near-neutrals - but there is no way to either put them or hold them in place. The near-neutral nature of beneficial mutations is strong evidence that every gene had to be designed, and that there is simply no conceivable way to build a gene one nucleotide at a time, via selection.

10. Putting bad mutations back in the picture. We have briefly considered a variety of powerful arguments about why progressive mutation/selection must be very limited in its scope. These arguments have temporarily excluded from consideration all deleterious mutations. However, in reality, progressive selection must occur in the real world, where deleterious mutations outweigh beneficial mutations by perhaps a million to one. To be honest, we must now re-introduce deleterious mutations.

a) Muller's Ratchet - As I have mentioned earlier, when we study the human genome, we see that large blocks of DNA have essentially no historical evidence of recombination (Gabriel et al. 2002, Tishkoff and Verrelli, 2003). Recombination appears to be primarily between genes rather than between nucleotides. So within any limited gene sequence there is essentially no recombination. Any such block of DNA that does not have recombination is subject to "Muller's ratchet" (Muller, 1964). This means that the good mutations and the bad mutations cannot be separated. Since we know that the bad mutations overwhelmingly outnumber the good, we can be certain that any such stretch of DNA must degenerate. The hordes of bad mutations will always drag the rare good mutations down with them. While we are waiting for a rare beneficial mutation, bad mutations are piling up throughout the region. Even if we could succeed in accumulating perhaps a hundred "good" mutations within a region, and were waiting for the next one to come along - we would start to see many of our good mutations start to back-mutate into the bad. Time is our enemy in this situation - the more time, the less information. Muller's ratchet will *kill* a new gene long before it can take shape.

b) <u>Too much selective cost</u> - In previous chapters we have discussed the *cost of selection.* Haldane's dilemma only considers progressive selection. But we can only afford to "fund" progressive selection for beneficial mutations after we have paid for all other reproductive costs - including all costs associated with eliminating bad mutations. As we have already seen, there are so many bad mutations we cannot afford even to pay just the reproductive cost of eliminating *them.* Since we cannot afford to stop degeneration - we obviously have nothing left over to fund progressive selection. There is just one way around this. In the short run, we *can* fund progressive selection for a very limited number of traits - if we borrow "selection dollars" from our long-term struggle against bad mutations. However, we need to understand that this means that any short-term progress in terms of specific beneficial mutations, is paid for by faster genomic degeneration in the long run.

c) <u>Non-random mutation</u> - As it turns out, mutations are not entirely random. Can this help us to create new genes? No, it makes our problem much worse! For example, we now know that some nucleotide positions are much more likely to mutate than others ("hotspots"), and that certain nucleotides are favored in substitutions. Mutational "hot spots" will give us the mutant we want sooner in that location, but while we then wait for the complementary mutations within the "cold spots", the hotspots will proceed to back-mutate again. We are forced to keep re-selecting our good mutations within the hot spots, while we wait for even the first good mutation to occur within the cold spots. This makes things worse, rather than better. The greater tendency to mutate to a certain nucleotide, (let's say *T*), will help us in positions where *T* is desired, but it will slow us down whenever *G, C, or A* is desired. Therefore, *75% of the time the bias toward T mutations will slow down progressive selection.* "Non-random mutation" sounds good from the point of view of building information, but unfortunately we are not talking about the non-randomness of *design* - rather we are talking about a type of non-randomness which (ironically) is antithetical to information building.

We have reviewed compelling evidence that even when ignoring deleterious mutations, mutation/selection cannot create a single gene - not within the human evolutionary timescale. When deleterious mutations are factored back in, we see that mutation/selection cannot create a single gene - ever. This is overwhelming evidence against the Primary Axiom. ***In my opinion this constitutes what is essentially a formal proof that the Primary Axiom is false.***

In conclusion, the genome must have been designed, and could not have evolved. Yet we all know that "micro-evolution" (adaptive selection) does happen, correct? How can this be? To use the terminology of our earlier chapters, mutations are the dings, scratches, and broken parts of life. Therefore, I believe most useful variation must be *designed*. When we see adaptive selection occurring, we are usually witnessing segregation and recombination of useful variants of genes and gene components - **which were designed to segregate and recombine in the first place**. We are not usually seeing the result of random mutations - which are consistently deleterious. Selection operates to eliminate the worst of mutations, while favoring the most desirable recombinants and segregants of designed variation. For example, a single human couple, if they contained designed and functional heterozygousity at only a tiny fraction of their nucleotides, would produce (via recombination and segregation) an essentially unlimited range of useful diversity. It is this type of designed diversity that natural selection can act upon most effectively. All such designed variants would be expected to be created within useful linkage groups, and would have originated at high allelic frequencies. For example, in the case of a single human couple, there could be only four initial sets of chromosomes - so all initial nucleotide frequencies would be at least 25%. Functional linkage groups and high allele frequency allow for very rapid responsiveness to selection, and thus rapid local adaptation. Like an ordered deck of cards, the net information in such a scenario would be greatest at the beginning, but diversity would be greatest only after many hands had been played out. Except at the beginning, no new information would be required.

Figure 12.

Poly-constrained information, and poly-constrained DNA. DNA, like word puns, word palidromes, and word puzzles, contains poly-functional letters, words, and phrases. Such sequences can only arise by very careful *design*. They cannot arise by mistake, and once they are created, they cannot be "mutated" to make them better. An excellent example is the painstakingly crafted poly-functional Latin phrase shown above, (see Ohno and Yomo, 1991). This ancient word puzzle (dating back to 79 AD) has a translation something like - THE SOWER NAMED AREPO HOLDS THE WORKING OF THE WHEELS. It reads the same, four different ways - left to right, or from up to down - or starting at the lower right, from down to up, or from right to left. Any single letter change in this system destroys all four messages simultaneously (all four of which happen to be the same, in this example). Similarly, a simple sentence palindrome would be: ABLE WAS I ERE I SAW ELBA, which reads the same forward or backwards. Any letter change destroys both messages. A simple example of a poly-functional word would be *LIVE*, which backwards is *EVIL*. To change *LIVE* to *HIVE* might be desirable, but it turns *EVIL* which has meaning, to *EVIH* - which is meaningless. So this dual-meaning word, like the other examples above, is poly-constrained, precisely because it is poly-functional.

Figure 13.

Irreducible complexity. In our red wagon example, the simplest improvement one might imagine would be some misspellings that would convert our red wagon into a blue tricycle. This seemingly small evolutionary step forward could never happen by chance because it requires creation of many new components, each of which represents "irreducible complexity". It is not hard to imagine a misspelling that would change the paint code, or cause a wheel to fall off. However, to make an operational tricycle that actually *works* (something selectable by our quality control agent), requires extensive re-working of most of the components. Just one new component – the pedal apparatus – illustrates this. Creation of an operational pedal apparatus for our little red wagon could not arise from a few misspellings. *Several entirely new chapters* in the manual would be required for manufacturing and assembling the various components of the pedal apparatus. But the new pedal apparatus would still not work – without a place for one to sit, and a place for one's legs! Very obviously, the corruption of a wagon's assembly manual via random misspellings (even with the help of our quality control agent) could never result in a shiny new blue tricycle. It could only lead to a run-down and broken wagon.

Is the downward curve real?

Newsflash –
All evidence points to human genetic degeneration.

Under Dr. Crow's most optimistic (and entirely unrealistic) selection model, we still see "backwards evolution", with a disastrous decay curve for the fitness values of all the individuals within a population (Figure 10b). The rate of decline in that illustration is very consistent with Dr. Crow's own estimate - that the fitness of the human population is now degenerating 1-2% per generation (Figure 4). When Crow's model is corrected to be more realistic, allowing for differences between individual mutations, I do not believe the downward trend will ever level off. But even if we do not correct Crow's model, it still shows genetic decay and declining fitness, rather than evolutionary progress.

The nature of information and the correctly formulated analogy of the genome as an instruction manual, help us see that the genome must degenerate. This common-sense insight is supported by information theory (Gitt, 1997). The very consistent nature of mutations to erode information helps us see that the genome must deteriorate. The high rate of human mutation indicates that man must be degenerating. The prohibitive cost of selecting for large numbers of mutations simultaneously indicates that man must be degenerating. The problems of near-neutral mutations, selection threshold, and selection interference, all indicate that man must be degenerating. Even realistic modeling and numerical simulation show that we are degenerating.

For decades biologists have argued on a philosophical level that the very special qualities of natural selection can essentially reverse the biological effects of the second law of thermodynamics. In this way, it has been argued, the degenerative effects of entropy in living systems can be negated – making life itself potentially immortal. However all of the analyses of this book contradict that philosophical assumption. Mutational **entropy** appears to be so strong within large genomes that selection can not reverse it. This makes eventual extinction of such genomes inevitable. I have termed this fundamental problem **Genetic Entropy**. Genetic Entropy is not a starting axiomatic position – rather it is a logical conclusion derived from careful analysis of how selection really operates.

If the genome must degenerate, then the Primary Axiom is wrong. It is not just implausible. It is not just unlikely. It is absolutely dead wrong. It is not just a false axiom. It is an unsupported and discredited hypothesis which can be confidently rejected. Mutation/selection cannot even stop the loss of genomic information - let alone *create* the genome! Why is this? It is because selection occurs on the level of the whole organism, and cannot stop the loss of information due to mutation, which is immeasurably complex, and is happening on the molecular level. **It is like trying to fix a computer with a hammer - the microscopic complexity of the computer makes the hammer largely irrelevant. Likewise, the microscopic complexity of genomic mutation makes selection on the level of the whole individual largely irrelevant.**

In our first chapter we considered a closely analogous scenario, wherein we tried to advance transportation technology. We proposed using a robotic, nearly-blind "scribe" who makes misspellings within instruction manuals at the beginning of a car-manufacturing plant. We then added a robotic, nearly-blind "judge" who does quality control at the other end of the assembly line. No humans *ever* enter the plant. We asked – "Will the cars coming off the assembly line progressively get better or worse?" Will the cars evolve into spaceships? Everything we have been considering should make it obvious - the cars will not evolve into spaceships - in fact they will become progressively inferior cars. The quality control robot can at best delay the inevitable failure of the business. Does more time help in this scenario? No, infinite time would just give us infinite certainty that the system would fail. But our factory has neither infinite time, nor infinite resources. Real life is like a business, and it must continually "pay all its costs" - or go out of business (die).

Despite massive amounts of "mental conditioning" of the public by our educational institutions, I believe most people can still instinctively see that the relentless accumulation of random misspellings within assembly manuals can not transform a car into a spaceship. Our quality control agent will simply **never, ever** see a deviant car that has a rocket engine - no matter how long he waits, nor how many misspellings occur in the manual. This is because of probability, and because of the problem of "irreducible complexity", as described by Behe (1996). Even if the judge **did** see a deviant car with a rocket engine - he would only reject it - because such a car with a rocket engine is still not yet a better means of transportation. Our judge (natural selection) has an I.Q. of zero, and has zero foresight, and has no concept of what a spaceship might be. Therefore he (it) has no conception of what deviations could make a car more "spaceship-like", nor would he (it) ever select such deviations. The only possible way he (it) could select toward a *spaceship* - would be by selecting for better *cars* - which is clearly paradoxical. Even if our judge **could** begin to select for more "spaceship-like" cars (most emphatically he cannot), it would take such an astronomically huge number of misspellings to create a functional spaceship - it would essentially take forever. But remember - our car factory cannot afford to take forever, it has to pay its bills today. In fact, those cars that might be more "spaceship-like" would likewise be less "car-like" - in other words they would be dysfunctional products, analogous to biological monstrosities. Bankruptcy is very obviously just around the corner for any such a commercial enterprise. **Careful analysis, on many levels, consistently reveals that the Primary Axiom is absolutely wrong.**

Would any one of you care to invest your life savings in a new company which had decided to use the mutation/selection manufacturing scheme? Its promoters say that it will be all robotic - no human agents will be needed at the plant - and they assure us that the plant will just keep making better and better products. Remember, we are not talking about a separately-funded research and development program - we are talking about the revenue-generating, boom or bust, assembly line! Any buyers? As for myself - I have been taken in by such schemes too many times in my life. With a healthy dose of skepticism, I would certainly dismiss such a scheme as a fraud. Despite glossy brochures, and VIP endorsements, I know any such scheme could only make deteriorating cars - and could never produce a revolutionary new spaceship. I have better places where I will invest my life and my life-savings.

If the Primary Axiom is wrong, then our basic understanding of life history is also wrong (see Appendix 5). If the genome is degenerating, then our species is not evolving, but is essentially "aging". There appears to be a close parallel between the aging of a species and the aging of an individual. Both seem to involve the progressive accumulation of mutations. Mutations accumulate either within those cell lines which give rise to our reproductive cells, or within those cell lines that give rise to our body cells. Either way, the misspellings accumulate until a threshold point is reached, wherein things rapidly start to fall apart, resulting in a distinct upper range for lifespan. Human life expectancy presently has an average of about 70 years, and a maximum near 120. However, when first cousins marry, their children have a reduction of life expectancy of nearly 10 years. Why is this? It is because inbreeding exposes the genetic mistakes within the genome (recessive mutations) that have not yet had time to "come to the surface". Inbreeding is like a sneak-preview, or foreshadowing, of where we are going genetically as a species. The reduced life expectancy of inbred children reflects the overall aging of the genome, and reveals the hidden reservoir of genetic damage (recessive mutation) that has been accumulating. If all this genetic damage were exposed suddenly (if we were all made perfectly inbred and homozygous) – it would be perfectly lethal – we all would be dead, our species would instantly become extinct.

Genetic damage results in aging, and aging shortens lifespan. This is true for the individual and for the population. Logically we should conclude that if all of this is true, then at some time in the past – there must have been a time when there was less genetic damage in the genome, and thus longer lives, and less deleterious effects from inbreeding. Is there any evidence of this?

The Bible records a limited time when people had extremely long lives, and when inbreeding was entirely benign. In fact, the life expectancies recorded in the book of Genesis seem unbelievable. According to the Bible, in the beginning, people routinely lived to be more than 900 years old! From where we stand now, that seems absurd. But our perspective, and our understanding, are so very limited! *We still do not know* why most mammals have a maximal lifespan of less than 20 - but for man it is about 120 years. Even chimps have a maximal life-expectancy which is less than half that of man. However, we *know at least this much - mutation is clearly implicated in aging.* So if there were initially no mutations - wouldn't you expect the maximal human age to be much longer? From this perspective, apart from mutations, human ages of hundreds of years would not be so crazy – they would be logical. Indeed - why would we die sooner?

A recent paper by a mathematician and a theologian presents some fascinating data (Holladay and Watt, 2001). Their paper analyzes the lifespan of early Biblical characters, versus years born after the patriarch Noah. This Biblical data (recorded thousands of years ago), clearly reveals an exponential decay curve. This curve can only be described as "biological". The calculated 'line of best fit' is exponential, and corresponds to the actual Biblical data very closely (correlation coefficient = .94). The actual formula for the decay curve that best fits the data was found to be: $y = 386.6835 (e^{-0.00462214x}) + 70.065$. I have done a similar analysis, as shown in Figure 14.

The unexpected regularity of the Biblical data is amazing. We are forced to conclude that the writer of Genesis either faithfully recorded an exponential decay of human lifespans, or the author fabricated the data using sophisticated mathematical modeling. To fabricate this data would have required an advanced knowledge of mathematics, as well as a strong desire to show exponential decay. But without knowledge of genetics (discovered in the 19th century), or mutation (discovered in the 20th century), why would the author of Genesis have wanted to show a biological decay curve? It does not seem reasonable to attribute this data to some elaborate "stone-age fraud". The most rational conclusion is that the data are real, and that human life expectancy was once hundreds of years - but has progressively declined to current values. The most obvious explanation for such declining lifespans, in light of all the above discussions, would be genetic degeneration due to mutation (the downward curve is especially steep in the early generations, suggesting that at that time there may have been a substantially elevated mutation rate).

In conclusion, we can see that even using Crow's most optimistic model, the downward curve for fitness is real, and in fact closely matches the downward curve for lifespans recorded in the book of Genesis. The *bad news* is that our species, like ourselves, is dying. The Primary Axiom is wrong. Information decays. Genomes decay. Life is not going up, up, up - it is going down, down, down. Selection does not create information, and at best can only slow its decay.

Information -
Genetic information must erode over time. The actual rate for *all types* of mutations may be more than 600 per person per generation, and so given a diploid genome size of 6 billion, *we should be losing about one ten millionth of our total information per generation*

(this number is not affected by what fraction of the genome is actually functional). To the extent that we are thinking linearly, this does not sound very shocking (after all, we might think that to lose 100% of the information would take ten million generations). However information is not linear. *A major computer program can fail completely due to a single error.* A very "robust" program can withstand multiple errors, but even the best designed programs cannot tolerate large numbers of errors. The genome appears to be a program so well designed that it can tolerate tens of thousands of errors. It is amazingly robust - unlike anything designed by man. But for all that, the genome is still not immune to failure due to error accumulation. In 300 generations (6,000 years), if the rate of loss was constant and at its current level, we would lose about .003% of our total information. This is *huge* - (90,000 errors) - yet given the extremely robust nature of the genome, it is *conceivable.* However, if we continued to lose information at this same rate for 300,000 generations (6 million years) we would lose 3% of all our information! This would represent 90 million errors! This is inconceivable. No program could still be functional.

Information theory clearly indicates that information and information systems arise only by intelligence, and are only preserved by intelligence (Gitt, 1997). Computers and computer programs do not arise spontaneously, they are painstakingly designed. Even computer viruses, contrary to the public's perception, do not arise spontaneously. They are painstakingly and maliciously designed. The emergence of the internet has created a vast experiment to see if information can organize itself. It does not. Everything which is happening on the internet – even the bad stuff – is designed.

Information's fundamental nature is to degenerate, and this reality is reflected all around us – from the whisper game, to gossip, to chains of command systems, to the routine crashing of our computer systems. We whimsically call all this "Murphy's Law". The reason our information systems do not degenerate even *more* rapidly is because of our elaborate, intelligently-designed systems to stabilize and preserve information. Yet even the best designed information systems, apart from intelligent maintenance and the *continual intervention* of intelligence, will always still eventually breakdown. Computers are typically junk within 5 years.

The genetic systems of life can be seen as intelligently designed information systems, and natural selection can be seen as an intelligently designed stabilizing mechanism. Even though these systems appear to be superbly designed they are still degenerating - apart from the intelligent and benign intervention of their designer.

What is the mystery of the genome? Its very existence is its mystery. Information and complexity which surpass human understanding are programmed into a space smaller than an invisible speck of dust. Mutation/selection cannot even begin to explain this. It should be very clear that our genome could not have arisen spontaneously. The only reasonable alternative to a spontaneous genome is a genome which arose by design. Isn't that an awesome mystery - one worthy of our contemplation?

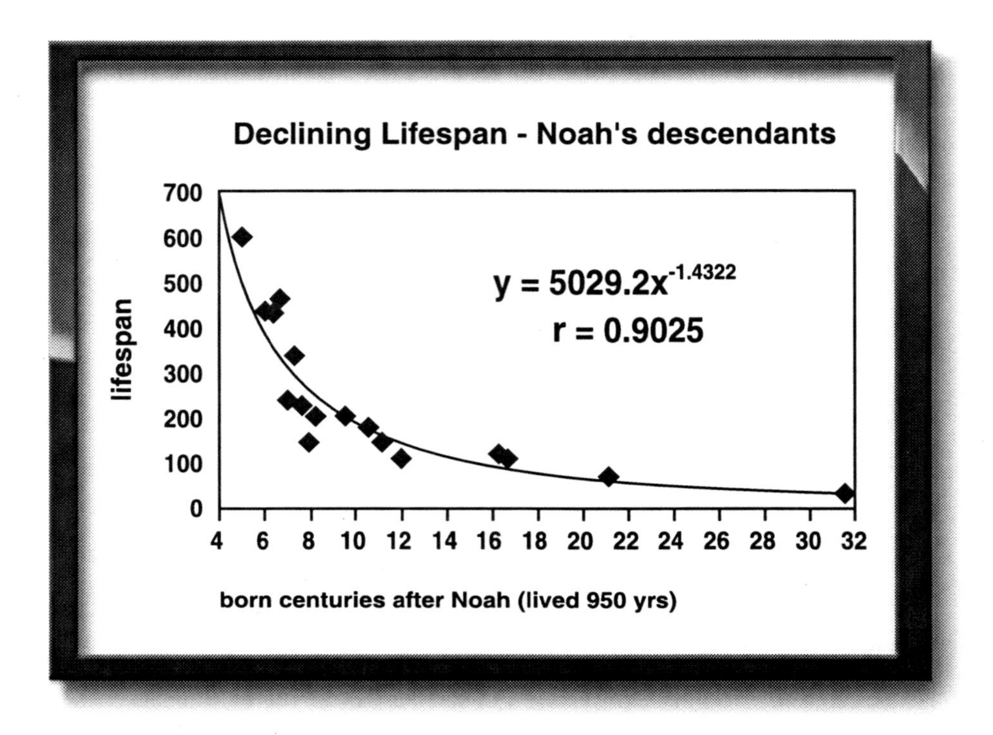

Figure 14.

Human life spans in early history. When Biblical life spans are plotted against time, for the generations after Noah, we see a dramatic decline in life expectancy, which has the strong appearance of a biological decay curve. Fitting the data to the "line of best fit" reveals an exponential curve following the formula: $y = 5029.2x^{-1.43}$. The curve fits the data very well - having a correlation coefficient of .90. It seems highly unlikely this Biblical data could have been fabricated. This curve is very consistent with the concept of genomic degeneration caused by mutation accumulation. The curve is very similar to the curves shown in Figures 4 and 10b, which are theoretical curves reflecting genomic degeneration.

Postlude

What hope?

Newsflash - There is a hope.

As you have so diligently stayed with me all the way through this book, and have now reached its end, perhaps you will not be offended if I diverge from what has been a strictly scientific discussion, and touch upon the philosophical. I would like to humbly put before you my own personal conclusion, regarding where our hope lies.

When I was still relatively young, I accepted the fact that I was going to die, and that all of the people I loved were going to die. I accepted it, but this realization certainly robbed me of joy – to say the least! But I was taught that there was still one hope – the world was getting better. Science was advancing. Culture was advancing. Even mankind was getting better. Through our efforts, we could make the world a better place. Through evolution, we could evolve into something better. Through *progress* - we might eventually defeat death itself. Perhaps we might someday even reverse the degeneration of the universe! So my personal hope was that I might in some small way contribute to such progress. I believe that this basic hope was shared, to a large extent, by my entire generation*.

I now believe this was a false hope. I still believe we should diligently apply ourselves to making this a "better world", and to be responsible stewards of the world which we have been given. But I see our efforts as a holding action - at best. While science can reasonably hope to prolong life, it cannot defeat death. Degeneration is certain. Our bodies, our species, our world – are all dying. It is simply not in our power to stop this very fundamental process. As we look around us – isn't this obvious? So where is the hope? If the human genome is irreversibly degenerating, then we must clearly look beyond evolution, in order to have a hope for the future.

One of my reviewers told me that the message of this book is both terrifying and depressing. He suggested that perhaps I am a little like a sadistic steward onboard the Titanic - gleefully spreading the news that the ship is sinking. But that is not correct - I hate the consequences of entropy (degeneration). I hate to see it in my own aging body, in the failing health of loved ones, or in the deformity of a new-born baby. I find it all absolutely ghastly, but also absolutely undeniable. Surely a real steward on the Titanic would have a responsibility to let people know that the ship was sinking - even if some people might hate him for it. I feel I am in that position. Responsible people should be grateful to know the *bad news*, so they can constructively respond to it. If we have been putting all our hope in a sinking ship, then isn't it necessary that we recognize this, and abandon our false hope? It is only in this light that we can appreciate bad news. Only in the light of the *bad news*, we can we really appreciate the *good news* – that there is a lifeboat.

Even as we cannot create life, we cannot defeat death. Yet I assert that there is One who *did* create life, and who designed the genome. I do not know how He did it, but somehow He surely made the hardware, and surely must have written the original software. He is called the Author of Life (Acts 3:15 – NIV). I believe the Author of Life has the power to defeat death and degeneration. I believe this is the ***good news***.

It is my personal belief that Jesus is our hope. I believe that apart from Him there is no hope. He gave us life in the first place - so He can give us new life today. He made heaven and earth in the first place - so He can make a *new* heaven and earth in the future. Because He rose from the dead - we can be raised from death, even that death which is already enveloping us. In these profound yet simple truths, I believe there is a true hope. I believe this is the hope which is unshakable, because I believe it is founded on the One who is eternal. It is a hope that has withstood the attacks of time, and the corruption of religion. It is a hope freely given, to anyone who would receive it today. I humbly put before you this alternative paradigm for your consideration - Jesus is our one true hope.

* *Kimura, 1976: "Shall we be content to preserve ourselves as a superb example of living fossils on this tiny speck of the universe? Or, shall we try with all our might, to improve ourselves to become supermen, and to still higher forms, to expand into the wider part of the universe, and to show that life after all is not a meaningless episode?".*

A deeply entrenched ideology

"It is obvious that the omnipotent power of natural selection can do all things, explain all things... ".

I have lost the source for the above statement, which came from an early Darwinist. But it does not matter. The ubiquitous nature of the philosophy which underlies this statement makes its source irrelevant. It could have come from just about any Darwinist - in fact just a few years ago I might have said it myself. More than 100 years after the above "statement of faith" was made, the Primary Axiom still captivates the minds and loyalties of most scientists. This is especially true among geneticists. However, when we look at numerous quotes from some of the most prominent population geneticists who ever lived, it appears that their commitment to the Primary Axiom is not based upon evidence, but is fundamentally an ideological commitment. Repeatedly, their own detailed analyses run counter to the Axiom. They seem to remain bound to the Axiom, not in light of their evidence, but in spite of their evidence. Hence they continuously need to explain why the Axiom can run counter to both common sense and their own data - and yet still must be considered axiomatically true.

All the following quotes are from leading evolutionary geneticists (except for Hoyle, who was a prominent physicist). I hold these scientists to be scientifically superior to myself - but I contend that they have built their work (perhaps their very lives?) upon a false axiom. I have extracted specific statements from their papers, which do not reflect their own philosophy, but which point to the problems I am raising.

Few university-based geneticists today (personally, I am semi-retired) would choose to openly discuss the weaknesses of the mutation/selection theory. Indeed, I suspect most geneticists have never even seriously considered such weaknesses (although I am aware of numerous skeptics who choose to remain "in the closet"). Most university geneticists have never seriously questioned the Primary Axiom. This is because by faith they have always accepted it as axiomatically true - even as I once did myself. So they have not even bothered examining the full extent of the problems - not with an open mind and an open heart. Every single study, every single paper, seems to be designed "to make the theory work". The ideological commitment to the Primary Axiom among geneticists is tremendous. However, among some (I think the ones most secure in their faith in the Axiom) there has been open acknowledgement of specific problems. These acknowledged specific problems, when combined, powerfully argue against the Primary Axiom. The following quotes illustrate this (all emphases below are mine).

Haldane's Dilemma

J. B. S. Haldane. 1957. The cost of natural selection. J. Genetics 55: 511-524.

"It is well known that breeders find difficulty in selecting simultaneously for all the qualities desired in a stock...***in this paper I shall try to make quantitative the fairly obvious statement that natural selection cannot occur with great intensity for a number of characters at once...***".

"I doubt if such high intensities of selection have been common in the course of evolution. I think n = 300 (300 generations), which would give I = 0.1 (10% total selective elimination in the population), is a more probable figure."

"If two species differ at 1,000 loci, and the mean rate of gene substitution, as has been suggested, is one per 300 generations, it will take at least 300,000 generations (6 million years)...."

"Even the geological time scale is too short for such processes to go on in respect to thousands of loci... can this slowness be avoided by selecting several genes at a time? I doubt it...."

"... the number of deaths needed to secure the substitution by natural selection of one gene ... is about 30 times the number of organisms in a generation...***the mean time taken for each gene substitution is about 300 generations.***"

"...***I am convinced that quantitative arguments of the kind put forward here should play a part in all future discussions of evolution."***

Haldane was the first to recognize there was a cost to selection which limited what it could realistically be expected to do. He did not fully realize that his thinking would create major problems for evolutionary theory. He calculated that in man, it would take 6 million years to fix just 1,000 mutations (assuming 20 years per generation). He could not then know that the number of actual genetic units is 3 billion, and that at least 1 million new mutations would be entering any hypothetical pre-human population each generation - most of which would require selection. Man and chimp differ by at least 150 million nucleotides representing at least 40 million hypothetical mutations (Britten, 2002). So if man evolved from a chimp-like creature, then during that process there were at least 20 million mutations fixed within the human lineage (40 million divided by 2), yet natural selection could only have selected for 1,000 of those. All the rest would have had to have been fixed by random drift - creating millions of nearly-neutral deleterious mutations. This would not just have made us inferior to our chimp-like ancestors - it would surely have killed us. Since Haldane's paper, there have been repeated efforts to sweep "Haldane's dilemma" under the rug, but the problem is still exactly the same. ReMine (1993, 2005) has extensively reviewed the problem, and has analyzed it using an entirely different mathematical formulation - but has obtained identical results.

Kimura's Quandary

Kimura, M. 1968. Evolutionary rate at the molecular level. Nature 217:624-626.

"... in the evolutionary history of mammals, nucleotide substitution has been so fast that, on average, one nucleotide pair has been substituted in the population roughly every two years. This figure is in sharp contrast to Haldane's well-known estimate ... a new allele may be substituted in a population every 300 generations...". " at the rate of one substitution every two years... the ***substitutional load becomes so large that no mammalian species could tolerate it...***". "This brings us to the ***rather surprising conclusion ... the mutation rate per generation for neutral mutations amounts to roughly ... four per zygote...".***

Kimura's estimate of the actual mutation rate was 25-100 fold too low. But it is amazing how easily evolutionary theorists can accommodate any new data - they seem to have an infinitely flexible model - allowing continuous and unlimited development/revision of their many scenarios.

Kimura, M. The Neutral Theory of Molecular Evolution. Cambridge University Press. p.27.

"This formula shows that as ***compared to Haldane's formula the cost is larger by about 2...*** under the assumption that the majority of mutation substitutions at the molecular level are carried out by positive selection***... to maintain the same population number and still carry out mutant substitutions ... each parent must leave ... 3.27 million offspring to survive and reproduce.*** This was the main argument I used when I presented the neutral mutation-random drift hypothesis of molecular evolution...".

Kimura realized that Haldane was correct, and that selection must occur extremely slowly, and that it can only affect a limited number of mutations simultaneously. Kimura also realized that all the evolutionists of his time were evoking too much selection for too many loci, leading to absurd costs (more than 3 million offspring selected away - for every adult!?). He developed his neutral theory in response to this overwhelming evolutionary problem. Paradoxically, his theory led him to believe that most mutations are un-selectable, and therefore most genetic information must be irrelevant, and most "evolution" must be independent of selection! Because he was totally committed to the Primary Axiom, Kimura apparently never considered the possibility that his cost arguments could most rationally be used to argue against the Axiom's very validity.

Muller's Fear

Muller, H.J. 1950. Our load of mutations. Amer. J. Human Genetics 2:111-176.

"It would mean an ever heaping up of mutant genes...degradation into utterly unrecognizable forms, differing chaotically from one individual of the population to another... it would in the end be far easier and more sensible to manufacture a complete man de novo, out of appropriately chosen raw materials, than to try to fashion into human form those pitiful relics which remained. For all of them would differ inordinately from one another, and each would present a whole series of most intricate research problems... if then the eliminated 20% failed involuntarily ...the remaining 80%, although they had contrived to reproduce would on the whole differ from the doomed filth but slightly... practically all of them would have been sure failures under primitive conditions... ***it is very difficult to estimate the rate of the degenerative genetic process...***". (pp. 146-7).

"... the open possibility that the deterioration consequent on the present relaxation of selection may after all be a good deal more rapid than has commonly been imagined... it is evident that the natural rate of mutation of man is so high, and his natural rate of reproduction so low, that not a great deal of margin is left for selection... if u has the minimal value of 0.1... an average reproductive rate of 2.4 children per individual would be necessary... without taking any account whatever of all the deaths and failures to reproduce for non-genetic causes***... it becomes perfectly evident that the present number of children per couple cannot be great enough to allow selection to keep pace with a mutation rate of 0.1...if, to make matters worse, u should be anything like as high as 0.5 ..., our present reproductive practices would be utterly out of line with human requirements.***" pp. 149-50.

Muller calculated that the human fertility rate of that time (1950) could not deal with a mutation rate of 0.1. Since that time, we have learned that the mutation rate is at least 1,000-fold higher than he thought. Furthermore, fertility rates have declined sharply since then.

"...the present genetic load is a serious one... increasing it by only 25%... would be a matter of grave consequence, while its doubling would be calamitous ***... if u should rise above 0.5, the amount of selective elimination required ... would, as we have seen, be greater than the rate of effective reproduction of even primitive man would have allowed... genetic composition would deteriorate continuously, while the population would meanwhile diminish in numbers, all the way to the point of disappearance."***

Muller concluded that if the mutation rate was as high as 0.5, early man could not have evolved - he would have degenerated and become extinct. But known mutation rates are 200-600 fold higher that this!

"... in many civilized nations, the birth rate is held down to an average of not much more than two per family, ***the upper mutation rate, that beyond which equilibrium is impossible, must be much lower than 0.5 and, as we have seen, perhaps lower than 0.1, even if selection were to be given full scope***." P. 155.

"If we now postulated that the conditions of raised mutation rate, low birth rate... it would be very problematic whether or not this decline would eventually be arrested." p. 156.

Muller states above, that even if we established the most severe selection scheme, modern fertility levels are not high enough to stop genetic deterioration - if mutation rates are as high as 0.5. Since then we have learned that our actual mutation rates are 100-300! Furthermore, fertility rates have sharply declined!

Muller's Ratchet

Muller, H.J. 1964. The relation of recombination to mutational advance. Mutation Research 1:2-9.

"There comes a level of advantage, however, that is too small to be effectively seized upon by selection, its voice being lost in the noise, so to speak. This level would necessarily differ greatly under different circumstances (genetic, ecological, etc.), but this is a subject which has as yet been subjected to little analysis ... although deserving of it."

Muller anticipated the problem of near-neutral mutations, but failed to see the profound problems they create for evolutionary theory. He did realize that many diverse circumstances (not just population size) would amplify this problem.

"It might appear as though a species without recombination would be ... subject to genetic degeneration... This might be thought to be the case ... However, this conclusion, which was misleadingly stated by the present author in a recent abstract, is only valid for the artificial conceptualization...nevertheless ***...an asexual population incorporates a kind of ratchet mechanism, such that ... lines become more heavily loaded with mutation.***"

How extremely reticent Muller was, to acknowledge the very problem which soon came to bear his name! Throughout his career, Muller had a deep concern for radiation-induced mutation and human genetic degeneration, and was a leading advocate of eugenics. However, even as he wanted to warn the public of the problem of genetic deterioration, he appeared to be extremely careful not to make any statements which might detract from the "certainty" of evolutionary theory (apparently this reflected where he placed his highest loyalty). His statements above, about the problem of genetic degeneration within asexual species (for which he became famous - Muller's Ratchet), are so cautiously worded, that one can hardly discern his message.

Muller argued that his "ratchet" was of only limited relevance - because he thought mutations were extremely rare. Therefore, he thought each mutation could be dealt with as a discrete and distinctly-selectable unit, being eliminated one at a time. We now know that mutations are numerous and diffuse, thwarting any possible "one-at-a-time" elimination mechanism.

If we combine Muller's recognition of near-neutral (i.e. un-selectable) mutations (above), with his recognition of mutational advance (above), we see that no selection system can stop "Muller's Ratchet" - even in sexual species. Even if selection could eliminate every mutation for numerous generations, eventually a new mutation would always "sneak through". In asexual species, once a mutation is fixed in a genome, there is no way to go back - to get rid of it. Hence the information can only degenerate - it is a unidirectional (ratcheted) process. Each new generation will have to have more mutations than the last, and so will be inferior to the last. Ironically, if we extend our analysis, we realize that this is not just true in asexual species, but is also true in sexual species. The "ratchet" works because a certain fraction of the deleterious mutations will always "sneak past" selection, and become fixed in the genome. These will still vastly outnumber any possible beneficial fixations. Selection cannot separate the few good from the many bad, because they are in large linkage blocks - they cannot be teased apart. Therefore each part of the genome (each linkage block) must individually degenerate due to Muller's Ratchet.

Neel's Realization

J. V. Neel et al. 1986. The rate with which spontaneous mutation alters the electrophoretic mobility of polypeptides. PNAS 83:389-393.

"...gamete rates for point mutations...on the order of 30 per generation... ***The implications of mutations of this magnitude for population genetics and evolutionary theory are profound.*** The first response of many population geneticists is to suggest that most of these occur in "silent" DNA and are of no real biological significance. Unfortunately for that line of reasoning ... the amount of silent DNA is steadily shrinking. ***The question of how our species accommodates such mutation rates is central to evolutionary thought."***

Kondrashov's Question

A. S. Kondrashov. 1995. Contamination of the genome by very slightly deleterious mutations: **Why have we not died 100 times over?** J. Theor. Biol. 175:583-594.

" Tachida(1990) concluded that VSDMs (very slight deleterious mutations) impairing only one function - its interaction with nucleosomes - may lead to too high a mutation load...".

"Lande (1994) and Lynch et al (1994) ...concluded... VSDMs can rapidly drive the population to extinction ...."

"Simultaneous selection against many mutations can lead to further decline of N_e and facilitate extinction (Li 1987; Lynch et al 1984)."
"I interpret the results in terms of the whole genome and show, in agreement with Tachida (1990), that VSDMs can cause too high a mutation load, even when $N_e = 10^6$-10^7.... conditions under which the load may be paradoxically high are quite realistic...".

"...the load can become excessive even when U<1... as my analysis suggests - contamination by VSDMs implies an excessive load, leading to stochastic mutation load paradox."

"...selection processes at different sites interfere with each other."

"...because the stochastic mutation load paradox appears real - it requires a resolution."

"Chetverikov (1926) assumed the mutational contamination of a species increases with time, leading perhaps to its eventual extinction."

"accumulation of VSDMs in a lineage ... acts like a timebomb... the existence of vertebrate lineages ... should be limited to 10^6-10^7 generations."

If Dr. Kondrashov would just believe his own data, he would conclude that the Primary Axiom is wrong, and that genomes must degenerate. Instead, he eventually appeals to "synergistic epistasis" to wave away the problems which he has so brilliantly characterized.

Kondrashov's Numbers

S. Kondrashov. 2002. Direct estimates of human per nucleotide mutation rates at 20 loci causing Mendelian diseases. Human Mutation 21:12-27.

"...the total number of new mutations per diploid human genome per generation is about 100 ... at least 10% of these are deleterious... analysis of human variability suggests that a normal person carries thousands of deleterious alleles...".

Since this paper, Dr. Kondrashov has indicated to me, by way of personal communication, that 100 was just a lower estimate, and that 300 is his upper estimate. He also indicated to me that he now believes up to 30% of the mutations may be deleterious. This means that from his perspective "U" (deleterious mutations per person per generation) would be 30-90. This is 100-fold higher than would have previously been considered possible. In the end, he dismisses the entire problem with "synergistic epistasis" and "truncation selection".

Nachman and Crowell's Paradox

M. W. Nachman and S.L. Crowell. 2000. Estimate of the mutation rate per nucleotide in humans. Genetics 156: 297-304.

"The human diploid genome ... about 175 new mutations per generation. ***The high deleterious mutation rate in humans presents a paradox. If mutations interact multiplicatively, the genetic load associated with such high U would be intolerable in species with a low rate of reproduction*** ... for U= 3, the average fitness is reduced to .05, or put differently, ***each female would need to produce 40 offspring for 2 to survive*** and maintain population size. This assumes that all mortality is due to selection...so ***the actual number of offspring required to maintain a constant population size is probably higher***."

According to Kondrashov (see above), U (new deleterious mutations per person) is actually 10-30 fold higher than these authors claim (they are assuming 97% of the genome is silent junk). Furthermore, we know that in reality, only a small fraction of total mortality can be attributed to selection. In spite of their unrealistic assumptions, these authors still acknowledge a fundamental problem. However, they eventually wave it all away by evoking "synergistic epistasis".

Walker/Keightley's Degeneration

A.Eyre-Walker and P. Keightley. 1999. High genomic deleterious mutation rates in Hominids. Nature 397:344-347.

"Under conservative assumptions, we estimate that an average of 4.2 amino-acid-altering mutations per diploid per generation have occurred in the human lineage...".

"...close to the upper limit tolerable by a species such as humans...a large number of slightly deleterious mutations may therefore have become fixed in hominid lineages***... it is difficult to explain how human populations could have survived...a high rate of deleterious mutation (U >>1) is paradoxical in a species with a low reproductive rate...*** If a significant fraction of new mutations is mildly deleterious, these may accumulate ... leading to gradual decline in fitness".

"...***deleterious mutations rate appears to be so high in humans and our close relatives that it is doubtful that such species could survive...".***

"...the level of constraint in hominid protein-coding sequences is very low, roughly half of all new non-synonymous mutations appear to have been accepted... if deleterious new mutations are accumulating at present, this could have damaging consequences for human health...".

These authors are still underestimating the extent of the mutation problem - they only consider the mutations within the protein-encoding portion of the genome. The functional genome is 10-30 fold larger than this. Even with these lower estimates, they acknowledge a fundamental problem, and conclude that many deleterious mutations have accumulated during human evolution, and are probably still accumulating. Doesn't this clearly demonstrate that we are in fact degenerating, not evolving?

Crow's Concerns

J.F. Crow. 1997. The high spontaneous mutation rate: is it a health risk? PNAS 94:8380-8386.

"...The overall impact of the mutation process must be deleterious... ***the typical mutation is very mild. It usually has no overt effect, but shows up as a small decease in viability or fertility...each mutation leads ultimately to one 'genetic death'*** ... would surely be an excessive load for the human population... so we have a problem."

"...there is a way out... by judicious choosing, several mutations may be picked off in the same victim... all individuals with more than a certain number of mutations are eliminated... ***of course.... natural selection does not line up individuals and remove all those with more than a certain number of mutations ....the unreality of this model kept me for many years from considering this as a way which the population deals with a high mutation rate***... although truncation selection is totally unrealistic, quasi-truncation selection is reasonable."

"It seems clear that for the past few centuries harmful mutations have been accumulating ... The decrease in viability from mutation accumulation is some 1-2% per generation... if war or famine force our descendants to a stone-age life they will have to be content with all the problems their stone-age ancestors had, plus mutations that have accumulated in the meantime...environmental improvements means that average survival and fertility are only slightly impaired by mutation...***I do regard mutation accumulation as a problem. It is something like the population bomb, but with a much longer fuse***."

Dr. Crow acknowledges the fundamental evolutionary problems created by the discovery of high mutation rates - but tries to dismiss them using a very unrealistic theoretical model involving a very artificial selection system based on "mutation count". (Whether or not this artificial selection scheme employs truncation or just quasi-truncation is just a matter of splitting hairs). He goes on to acknowledge that humanity must now be genetically inferior to our stone-age ancestors - an amazing confession about the reality of genomic degeneration.

Dr. Crow also comments on the inherently deleterious nature of mutations:

"Even if we didn't have a great deal of data on this point, we could still be quite sure on theoretical grounds that mutations would usually be detrimental. For a mutation is a random change of a highly organized, reasonably smoothly functioning living body. A random change in the highly integrated system of chemical processes which constitute life is almost certain to impair it - just as a random interchange of connections in a television set is not likely to improve the picture." James Crow – "Genetic effects of radiation" - Bulletin of the Atomic Scientists vol. 14:19-20. (Jan. 1958).

Lynch's (et al.) Mutation Meltdown

M. Lynch, J. Conery, and R. Burger. 1995. Mutation accumulation and the extinction of small populations. The American Naturalist 146:489-518.

"As deleterious mutations accumulate by fixation, there is a gradual decline in the mean viability of individuals ...net reproductive rate is less... precipitates ***a synergistic interaction between random genetic drift and mutation accumulation, which we refer to as mutational meltdown...the length of the meltdown phase is generally quite short."***

"These results suggest ***that for genetic reasons alone, sexual populations with effective population sizes smaller than 100 individuals are unlikely to persist for more than a few hundred generations***, especially if the fecundity is relatively low."

"...our results provide no evidence for the existence of a threshold population size beyond which a population is completely invulnerable to a mutational meltdown...".

"...simultaneously segregating mutations interfere with each others' elimination by natural selection ...".

Lynch et al. understand that genetic degeneration is a major factor for all currently endangered species. They fail to go on to explain that if this is true, then the same would be true for most past extinctions, and that the basic process of extinction via genomic degeneration should logically apply to all higher genomes. They also acknowledge the problem of selection interference.

Higgins and Lynch - More Meltdown

K. Higgins and M. Lynch. 2001. Metapopulation extinction caused by mutation accumulation. PNAS 98: 2928-2933.

"Here we show that metapopulation structure, habitat loss or fragmentation, and environmental stochasticity ***can be expected to greatly accelerate the accumulation of mildly deleterious mutations... to such a degree that even large metapopulations may be at risk of extinction.***"

"...mildly deleterious mutations may create considerably larger mutational load ... because individually they are nearly invisible to natural selection, although causing appreciable cumulative reduction in population viability".

"***We find that the accumulation of new mildly deleterious mutations fundamentally alters the scaling of extinction time... causing the extinction of populations that would be deemed safe on the basis of demography alone.***"

"Under synchronous environmental fluctuations, the acceleration of extinction caused by mutation accumulation is striking... without mutation, extinction is 2,000 generations... ***with mutation accumulation the extinction time is just slightly longer than 100 generations...***"

"For a metapopulation in unfragmented habitat, mildly deleterious mutations are more damaging than highly deleterious mutations... just as in the case with large carrying capacity, ***the mild mutational effects are the most damaging, causing minimal time to extinction.***"

"Early work suggested...accumulation of deleterious mutations may threaten small isolated populations ... here we show that accumulation of deleterious mutations may also be a significant threat to large metapopulations ... ***the decline is sudden, but extinction itself still takes a while to occur, the metapopulation may be completely inviable on intermediate or long time scales, although appearing healthy on short time scales.***"

Higgins and Lynch make a strong case for genomic degeneration as a general problem for all mammals, and all similar animal populations. They point out that even genetic damage which has already happened and ensures eventual extinction – still takes time to take effect. In the meantime, the species can still appear healthy and viable. Is that possible in the case of man? Human fertility and human sperm counts are both now dramatically declining (Carlsen et al., 1992). Many nations are now facing negative population growth – probably due to non-genetic causes. But is it not conceivable we could be in the early stages of mutational meltdown?

Hoyle's Big Picture

Fred Hoyle. 1999. Mathematics of Evolution. Acorn Enterprises, LLC, Memphis, TN. (note - unlike the others quoted here, Dr. Hoyle was not a geneticist, but a highly distinguished theoretical mathematician and physicist).

"The aging process shows, indeed, that ***statements one frequently hears, to the effect that the Darwinian theory is as obvious as the Earth going round the Sun, are either expressions of almost incredible naiveté or they are deceptions...with such widespread evidence of senescence in the world around us, it still seems amazing that so many people think it "obvious"*** that the biological system as a whole should be headed in the opposite direction...".

"The best natural selection can do, subject to a specific environment, is hold the deleterious mutations in check. ***When the environment is not fixed there is slow genetic erosion, however, which natural selection cannot prevent.***"

"... natural selection cannot turn back deleterious mutations if they are small, and over a long time a large number of small disadvantages escalate to a serious handicap. ***This long term inability of natural selection to preserve the integrity of genetic material sets a limit to its useful life...***".

Howell's Challenge

Howell et al. 1996. Evolution of human mtDNA. How rapid does the human mitochondrial genome evolve? A. J. Hum. Genet. 59: 501-509.

"We should increase our attention to the broader question of how (or whether) organisms can tolerate, in the sense of evolution, a genetic system with such a high mutational burden."

Howell's challenge to us is based upon his own data, which suggested that the mutation rate just within the mitochondrial genome might be approaching one mutation per person per generation. He is right—just 0.1-1.0 mitochondrial mutations per person create insurmountable problems for evolutionary theory. Yet this is nothing, compared to the hundreds of mutations simultaneously occurring within the other chromosomes.

How many nucleotides can be under selection at one time?

How many traits or nucleotides can be selected for, simultaneously, in a given breeding population? This is a very important question, and one that has not really been addressed adequately before. It is relevant to artificial breeding, and is relevant to population biology and evolutionary theory. The question can be dealt with on a mathematical basis.

Definition of C and c

Total selective cost (C) to a population is that fraction of the population that is not allowed to reproduce - in order to achieve all selection. On the simplest level, we will assume that the fraction of the population which is selected against is C, and produces zero offspring. So the remainder (1-C) is that part of the population which is selected for, and is allowed to reproduce at the "normal rate". Different species can afford different selective cost. For example, a plant species may produce one hundred seeds per plant. Such a species can afford to have a C value of .99. This means that 99% of the seedlings can be selected away and the population can still reproduce fully. In the case of man, the current human fertility rate is now roughly 3 children for every 2 adults. So in the human population, maximally, only one child of the three can be selected away - while still maintaining population size. So in mankind, selective cost must be below 1/3 of the population, and C must not exceed .33. In reality, even this cost is much too high. This is because there are many individuals who fail to reproduce for non-genetic reasons (accidental death, personal choice, etc.). We cannot know how often failure to reproduce is due to non-genetic effects - but it is surely very large. So a realistic estimate of allowable selective cost in mankind must be less than 25%, probably near 10%. To be generous, we may assume C might be as high as .25. To determine upper theoretical limits for man, we may assume an unrealistically fertile human population wherein C = .50 (half of all children are eliminated from the breeding population for genetic reasons every generation).

Cost per trait (c) is that part of the population which is eliminated to affect a specific trait (or nucleotide). If we are selecting against a given trait (or nucleotide), we need to decide how strongly we will select against it. In other words, how much of the total population are we willing to eliminate to improve that trait? The part of the population that is eliminated for that trait is the selective cost (c) for that trait, and represents the "selection pressure" for that trait. For example, if 10% of the population is eliminated to affect a given trait, then for that trait, c = .10. If c = .01, then 1% of a population is prevented from reproducing in order to affect that trait.

Additive model

The simplest model to understand is the additive model. In this model, we assume that selection is additive, and that selection for all traits is implemented simultaneously. For example, if we could afford to eliminate 25 individuals from a population of 100, we could simultaneously eliminate one individual to affect one trait - and so we could affect 25 different traits (or 25 nucleotides). The general formulation would be as follows. Total population cost (C), would be the sum of all costs for each trait or nucleotide (c). So $C = c_1 + c_2 + c_3 \ldots c_n$, where "n" is the number of traits. Assuming that the selection pressure on each trait is the same, then C = n x c. In the case where selection pressure per trait is .001 (1 individual is eliminated out of 1000, to affect a given trait), and where total cost of selection is limited to 25% of the population, then .25 = n x .001. So in this instance, the maximal number of traits that can be selected for is 250. However, in such a case, even though 250 traits could be under selection, the selection pressure per trait would be vanishingly small, resulting in little or no selective progress over time. *Selective progress* approaches zero very rapidly, as more and more traits are put under selection (see Figure 6a - Chapter 5).

Multiplicative model

The multiplicative model is slightly more realistic and more complex than the additive model. In this model there is first selection for one trait (or nucleotide), and then what is left of the population is subjected to selection for the next trait (or nucleotide). So selection is sequential rather than simultaneous. After one round of selection, the remainder of the population is mathematically, 1-c. If there are two traits one wishes to select for, then one multiplies the remainder of each: (1-c) x (1-c) and then subtract from one - to see the total cost of selection. For example if we eliminate 10% of a population for one trait and then 10% of the remainder for another trait, then our total cost is: 1 - [(1-.1) x (1-.1)] = .19. In other words, 81% of the population remains to reproduce after selection for these two traits. For many (n) traits under selection, assuming that each trait undergoes approximately the same selection intensity, the equation can be generalized as follows: $C = 1 - (1-c)^n$.

In Figure 6b, Chapter 5, I have plotted the number of traits under selection against the maximal allowable selection intensity per trait, assuming a multiplicative model and a realistic value for C = .25 (25% of a human population can be eliminated for all selective purposes). As can be seen, the shape of this curve is essentially identical to the additive model. As the number of traits under selection increases, the allowable selection pressure per trait falls off exponentially, rapidly approaching zero. This basic pattern does not change even where the population is extremely fertile (Figure 6c - Chapter 5). Even if we could assume an exceedingly fertile human population (C = 0.5), allowable selection per trait falls off extremely rapidly as "n" increases. Even when considering an extremely fertile species, such as a seed-producing plant, wherein C might be as high as .99, maximal allowable selection pressures become very small when there are more than 1000 traits (nucleotides) under selection.

What do these vanishingly small selection pressures mean? As the selection pressure for a trait approaches zero, selective progress also approaches zero, and the time to alter a trait via selection approaches infinity. As selective progress tends toward being infinitely small and infinitely slow, we realize we have a problem. This is because new mutations are constantly flooding into a population at high rates. We do not have "deep time" to remedy our degeneration problem. We need to eliminate mutations just as fast as they arise, or mutations get "imbedded" in the population (due to drift and fixation). Even more significantly, as the allowable selection pressures get very small, at some point effective selection truly *stops*. This is because of the phenomenon of "noise" and genetic drift. *A point will always be reached where selection halts altogether* - depending on population size and the total amount of biological "noise" associated with reproduction.

The exact threshold where selection completely breaks down is somewhat "fuzzy". However, some common sense can help put this issue in perspective. A selection system for a given trait, which cannot even remove one individual from a breeding population of 1000, is certainly suspect! This corresponds to a selection cost for that trait of .001. Given the high level of "noise" within human populations, when the selection cost is less that .001, effective selection for that trait may cease entirely. Another way of saying this is as follows: in a population of 1000 people, if we are not allowed to remove even one (whole) person to affect a given trait, selection for that trait has effectively stopped and random drift is probably operational. Using this cut off point of .001, and the additive model, we can calculate that we can maximally select for only 500 traits (nucleotides) in a realistic human population, and only 990 traits in an idealized (extremely fertile) human population* (see Table 2). Yet what we know about human mutation rates indicates that we need to select for millions of nucleotide positions every generation, in order to stop genomic degeneration.

** Kimura alludes to the same problem. Even though he does not show his calculations, he states that only **138** sites can be selected simultaneously when C=**.50**, and s (or 'c') =**.01** (Kimura, **1983**, p. **30**).*

How many genic units can be selected simultaneously (assuming a minumum for c = .001)?

C	n^a	n^m
.25	250	300
.50	500	700
.99	990	4,600

n^a = The maximal number of genic traits that can be selected under an additive model.

n^m = The maximal number of genic traits that can be selected under a multiplicative model.

Table 2.

How many nucleotides can be selected simultaneously?

The phenomenon of Unity, and the concept of Integrated Complexity.

The puzzle of how to recognize Intelligent Design has been gradually coming together, over a long period of time. There has always been an intuitive recognition of design in nature, which is thus the logical default perspective. To the extent that some people wish to reject the obvious, design was later explicitly proclaimed through Scriptural Revelation (Genesis through Revelation). Still later, design was argued by essentially all of the "Founding Fathers" of science, including Copernicus, Bacon, Newton, Pasteur, Maxwell, Faraday, and Kelvin. Paley (1802) was the first to put forward the argument of *complexity* as evidence of design. This concept has more recently been refined by Behe (1996), into the argument of *irreducible complexity*. The complexity argument has been further elaborated into the two related arguments – that of *information theory* (Gitt, 1997), and *specified complexity* (Demski, 1998). However, I believe there is still at least one more useful diagnostic of design, which has not yet been fully described. This is the phenomenon of Unity, which arises as a result of *integrated complexity*.

One diagnostic of design is the *comprehensive integration of a large numbers of components* – which is what I am calling **Integrated Complexity**. Integrated Complexity underlies the natural phenomenon which we recognize as **Unity**. The phenomenon of unity is an objective reality. Unity is readily recognized by any rational person, and is therefore not merely subjective. Unity is therefore a legitimate subject of scientific analysis. Unity arises through the *comprehensive integration* of very many parts. A jigsaw puzzle has unity. A pile of sand does not.

A fighter jet is made up of thousands of component parts and countless atoms, but it has **Unity** of both function and form. This is what makes it readily recognizable as a product of design. It exists as a single integrated unit, above and beyond all its components. In its original un-degenerate state, every single component has a purpose and a place, and each part is perfectly integrated with all the rest. Despite its countless components, the jet exists in a non-plural state. This is the essence of the term "unity" (oneness). We do not say: "oh, look at all those pieces of metal and plastic" – we say: "oh, look at that plane". It is not even remotely adequate to say that a plane is more than the sum of its parts. A jet is a new reality existing on a totally different level than any of its parts. It can fly – the parts can not. In a similar manner, it is foolish to say that a spaceship is more than lots of metal. It is also foolish to say that life is more than the sum of its parts. These are all obscene understatements. We might as well state that there is more than one drop of water in the sea. These things are so grossly obvious - how can we justify even saying them out loud – unless we are talking to someone who is in a trance?

A human being contains over 100 trillion cells. But we **are not** 100 trillion cells. I repeat - that is not what we are. We are each truly a **singular entity**, united in form and function and being. We are the nearly perfect integration of countless components – and as such we comprise a singular new level of reality. The separateness of our existence as people – apart from our molecules – is both wonderfully profound, yet childishly obvious. Only a deep spiritual sleep could blind us to this reality. We desperately need to wake up. When we awake to the reality of unity, we also awake to the reality of *beauty*. We begin to realize that what we call "beauty" is simply the recognition of the comprehensive unity of designed things. In this light, beauty is not merely subjective. In this new light, beauty, like unity, can be seen as a truly objective and concrete reality*.

In their more poetic moments, scientists sometimes refer to the beauty of unity as *elegance*. Elegance is design that is so excellent and wonderful

**As a personal aside, the converse of beauty is ugliness. I would like to suggest that ugliness is also an objective reality. Ugliness is the corruption of design, and the marring of unity. This is why a tumor or even a wart can objectively be considered ugly. It is why aging is an "uglifying" process. It is why rusting cars, rotting homes, biological deformity, broken families, wars, lies - are all truly ugly.*

that every detail, every aspect, comes together perfectly - to define something new - **a comprehensively integrated whole**. Unity can be seen as the startling absence of loose ends or frayed edges. For example, in man, every cell has its place and function - in such a way as to specify **wholeness**. The human profile, like the profile of a sleek jet airplane, proclaims elegance of form and unity of purpose. I would like to submit to you that unity is the concrete and objective basis for what we call beauty. I believe it is also an unmistakable diagnostic for very high level design.

The amazing unity of a human body (our phenome) should be obvious to any thoughtful person who is even remotely acquainted with biology. When we see a human being, we do not think: "look at all those cells and tissues". We see a single entity – a person.

What does all this suggest, regarding the human genome? The genome is the presumed basis, underlying the phenome's unity. Yet amazingly, most modern geneticists see the genome as being essentially a dis-united pile of nucleotides. All our collective genomes are said to merely comprise a "pool of genes". This is the very antithesis of unity. The genome is seen as a vast array of molecules, which is largely accidental and almost entirely random. Each nucleotide supposedly arose and is "evolving" (or drifting) independently. This entire pattern of thought (i.e. man is just a bag of molecules) is termed *reductionism*. The typical modern geneticist sees the genome as primarily "junk DNA", within which are imbedded over a million parasitic "selfish genes" (they also acknowledge that there are some real bits of information - a few tens-of-thousands of functional genes). It is widely assumed that each selfish gene has its own selfish agenda - propagating itself at the expense of the whole.

Rationally, how could this all be true? In light of the second law of thermodynamics, does it seem possible that the phenome's amazing unity and order, arises entirely from a fragmented and chaotic genome? Rationally, if the order and unity of the phenome derives from the genome, then shouldn't the genome be *more complex* and *more integrated* than the phenome?

Imagine entering the intergalactic starship, the S.S. Phenome. You go past doors labeled "Warp Speed Engine Room" and "Holodeck". Then you see a door marked "Office of the Senior Architect and the Chief Engineer". You open the door and you see an office that is a complete wreck. Papers are strewn everywhere, there is the smell of rotting food, and computer screens are broken. Standing on a desk are two chimpanzees – fighting for a banana. Would you be so naïve as to believe that you were actually looking at the Senior Architect and the Chief Engineer of the S.S. Phenome? Would you actually think the S.S. Phenome could have been *created* and *maintained* through this office – in its degenerate and chaotic condition? Yet this is really the modern view of the genome! This is the ruling paradigm regarding the genome's very nature, and describes the *idiot master- genius slave* relationship of genome and phenome. In this light, shouldn't we be critically re-evaluating our view of the genome? Isn't it time for a paradigm shift?

If Integrated Complexity is actually diagnostic of design, and if the genome really was originally designed, then we would predict that the genome should show extensive evidence of integration and unity. We should be able to discover many levels of unity of form and function within the genome. I believe this is now beginning to happen. I predict that this will be seen more and more in the coming years, as we unravel the many elaborate and multi-dimensional patterns within the genome. I predict that when we understand the genome better, we will see integration and unity - at all levels. But I also predict that we will see more and more evidence of degeneration and corruption of the original design - since mutation is degenerative and selection can not prevent mutational degeneration. The genome is clearly experiencing an enormous amount of change due to our high rates of mutation. But it seems to me what we are seeing is entirely "downhill" change. Such random change can't possibly be the origin of Integrated Complexity. Unity (comprehensively integrated complexity) simply can not be built one mistake at a time (as the main body of this book clearly demonstrates).

The profound unity of life exposes reductionism for what it truly is - a type of spiritual blindness. Reductionism is simply a profound ignorance of the unity which is self-evident all around us. More specifically, the Primary Axiom, with its "gene pools" and independent evolution of individual nucleotides, is merely extreme reductionism applied to biology. It is thus inherently invalid. In a sense, this makes all the arguments of this book unnecessary. It is my personal conviction, that even apart from all of the genetic arguments of this book, the Primary Axiom is fully invalided, simply by the all-pervading reality of the phenomenon of Unity.

Can Gene Duplication and Polyploidy Increase Genetic Information?

In opposition to the main thesis of this book, some would like to argue that duplication is the key to understanding how genetic information can increase spontaneously. It is certainly true that duplications occur spontaneously within all genetic systems. Duplication is a form of mutation, and the size of a duplication can be very small (one or just a few nucleotides) or very large (one chromosome or all chromosomes together are duplicated). When one chromosome is doubled it is called *aneuploidy*, when all chromosomes are doubled, this is called *polyploidy*. As with word-processing errors, a single letter can be duplicated, a single word can be duplicated, a whole chapter can be duplicated, a whole book can be duplicated, or the whole library can be duplicated. The question is this – "Do such duplications create new information?"

Iff I repeaaat a llettter, does it immmprove my sentenccce? If I repeat my sentence, do I tell you more? If I repeat my sentence, do I tell you more? If I repeat my sentence, do I tell you more? If this page occurred a second time elsewhere in this book, would the book be better? If every page of this book was written in duplicate, would you learn twice as much from it? Obviously, all these types of duplications are deleterious, regardless of the scale. They do not increase communication, they obviously disrupt it. How could anyone think this type of duplication is a realistic method for the spontaneous amplification of useful information? The answer is, of course, that such people imagine combining mutational duplication with almighty selection. But we have just dedicated most of this book to showing that while selection can slow mutational loss of information, it can not stop it. Most emphatically selection can not reverse this loss. It should be obvious by now, if you have read this book, that nearly all duplications will be both deleterious and nearly-neutral – like all other classes of mutation. This means selection will only be able to eliminate the very worst duplications, the rest will relentlessly accumulate and gradually destroy the genome.

Does biological observation support this common sense view of duplication? It most emphatically does! Let us consider the human population. Are there any polyploidy humans? Of course not, duplicating all the human genome is absolutely lethal. Are there any aneuploid humans? Yes there are – a significant number of people have one extra copy of one chromosome. Do these individuals have more information? Most emphatically, they do not. While *aneoploidy* is entirely lethal for larger chromosomes, an extra copy of the smallest human chromosomes is not always lethal. Tragically, the individuals who have this type of "extra information" display severe genetic abnormalities. The most common example of this is Down's Syndrome – resulting from an extra copy of chromosome 21. There are countless smaller duplications and insertions which also have been shown to cause genetic disease. It should be obvious by now, if you have read this book, that most duplications will be deleterious and nearly-neutral – like all other classes of mutation.

It is widely recognized that duplication, whether within a written text, or within the living genome, destroys information. Rare exceptions may be found where a duplication in a text or within a genome is beneficial in some minor way (possibly resulting in some "fine tuning"), but this does not change the fact that overwhelmingly, random duplications destroy information. In this respect, duplications are just like the other types of mutations.

After a given gene has been accumulating deleterious mutations for a long time - it is in a partially degenerated state. If that gene is then duplicated, the deleterious mutations are duplicated with it. Does such duplication in any way slow down the degeneration process? Obviously not! Upon careful consideration, we can see that once there is a duplicate copy of a gene, both copies will degenerate faster than before. Why is this? It is because each would then have a "back-up copy", so selection will become relaxed for both copies. It is often claimed that after a gene duplication, one gene copy might then stay unchanged, while the other might be free to "evolve a new function". But neither of these events is actually feasible. Both copies will degenerate at approximately equal rates, due to the accumulation of near-neutrals, as we have been learning. Neither can stay unchanged. Furthermore, gene conversion should theoretically be continuously cross-contaminating both the reputed "unchanging copy" and the reputed "evolving copy". Gene conversions should theoretically also allow mutations within each gene to jump into the other copy – which should effectively increase the mutation rate for both copies. This will clearly also accelerate degeneration. In summary, duplicate genes should clearly contribute to each others accelerated degeneration, due to relaxed selection and accelerated mutation accumulation, and due to mutation scrambling via gene conversion. As if this is not enough, chapter 9 of this book clearly shows how unreasonable is the speculation that one gene copy is likely to "evolve a new function", even while both copies are irrevocably degenerating.

What about polyploidy plants? It has been claimed that since some plants are polyploidy (having double the normal chromosome number), this proves that duplication must be beneficial and must increase information. Polyploidy was my special area of study during my Ph.D. thesis. Interestingly, it makes a great deal of difference how a polyploid arises. If somatic (body) cells are treated with the chemical called colchicine, cell division is disrupted, resulting in chromosome doubling - but no new information arises. The plants that result are almost always very stunted, morphologically distorted, and generally sterile. The reason for this should be obvious - the plants must waste twice as much energy to make twice as much DNA, but with no new genetic information! The nucleus is also roughly twice as large, disrupting proper cell shape and cell size. In fact, the plants actually have less information than before, because a great deal of the information which controls gene regulation depends on gene dosage (copy number). Loss of regulatory control is loss of information. This is really the same reason why an extra chromosome causes Down's Syndrome. Thousands of genes become improperly regulated, because of the extra genic copies.

If somatic polyploidization is consistently deleterious, why are there any polyploidy plants at all – such as potatoes? The reason is that polyploidy can arise by a different process – which is called sexual polyploidization. This happens when an unreduced sperm unites with an unreduced egg. In this special case, all of the information within the two parents is combined into the offspring, and there can be a net gain in information within that single individual. But there is no more total information within the population. The information within the two parents was simply pooled. In such a case we are seeing pooling of information, but not any new information.

In a diploid, there can be two versions of the same gene. In such a heterozygous diploid, if one gene version is a dysfunctional mutant, and the other is a functional non-mutant gene – the later can act as a back-up copy of the former. Diploidy can thus be seen as a designed back-up system – designed in anticipation of the mutation problem (on the other hand, evolution can not anticipate anything, and so we can very reasonably conclude, should never produce any back-up systems). Sexual polyploidy essentially doubles potential heterozygousity, so there can be up to four versions of the same gene within the same individual. Such a system is thus doubly backed-up. Like the four redundant computers used on the space shuttle, there can be up to three mutant alleles at a given locus, but as long as the fourth is still functional – the plant is all right. So polyploidy does not provide a way to increase new information, but rather illustrates the importance of gene redundancy as a back-up system – which effectively slows down degeneration. The cost of such back-up systems is that selection can not remove mutations nearly as efficiently - so long term degeneration is even more certain. In some special cases, the extra level of gene-backup within a polyploidy can outweigh the problems of disrupted gene regulation and reduced fertility – and so can result in a type of "net gain". But such a "net gain" is more accurately described as a net reduction in the rate of degeneration.

What about duplicate genes and gene families? If having multiple gene versions can explain the utility of diploidy and polyploidy, it can likewise explain the utility of redundant copies of a given gene at different locations within the genome. Normally, when a redundant version of a gene is seen within another part of the genome, it is simply assumed by theorists that it must have arisen by "an ancient gene duplication". They often add the general presumption of subsequent "mutational divergence". But this is all based upon theoretical inferences, not observation. If a gene is redundant within the genome, such redundancy could just as logically be understood as having a designed function – such as gene back-up or complex gene regulation.

The simple-minded notion that merely duplicating a gene might be beneficial is biologically naïve. Yes - it is possible that a gene duplication might increase that gene's expression. In fact, this is sometimes seen.

But simply increasing a gene's expression is usually deleterious (gene expression must be precisely regulated by elaborate and finely tuned molecular systems). Furthermore, duplication is a remarkably inefficient way to achieve such increased gene expression – how could evolution consistently be so inefficient?

Lastly, actual gene duplications not only mess up their own expression, they routinely mess up the expression of other genes. Much of my own career was spent in the production of genetically engineered plants. Industry and academia spent over a billion dollars in this endeavor. What was quickly discovered was that multiple gene insertions consistently gave lower levels of expression than single gene insertions. Furthermore, the multiple insertions were consistently less stable in their expression (can you start to see that gene regulation is very complex?). Additionally, a large percentage of all transgenic plants displayed other genetic defects – due to the disruptive effect of the extra DNA being randomly inserted into specific locations within the genome. Since the genome has a functional and highly specific architecture, any duplication or insertion should logically tend to disrupt that architecture. This is exactly what plant geneticists have been seeing.

The notion of gene duplication as a way to "evolve new information" has become very firmly entrenched within the evolutionary community. I believe this is partly because - "it must to be true – how else could evolution have happened?" I also believe that when a mantra is mouthed often enough - it takes on the appearance of Unassailable Truth. But careful analysis of what information really is, and how it arises, combined with a healthy dose of common sense, should reveal to us that random duplications are consistently bad. It is my personal opinion that "evolution through random duplications" is for the most part a widely-held *philosophical assumption*, rather than a scientifically-defensible observation. I believe that while it sounds quite sophisticated and respectable, it does not withstand honest and critical assessment.

Three Possible Objections

There are three possible objections to the thesis of this book, which I would like to address in this appendix. These issues are not dealt with in the body of this book because they are for more advanced readers, and would detract from the general readability of the main text.

Objection #1 – Mega-beneficial Mutations.

If there were occasional rare mutations which had a profoundly beneficial effect, then such mutations might outweigh all the harmful effects of the accumulating deleterious mutations. This might halt degeneration. For example, perhaps a single nucleotide substitution might increase the information content of the genome by 1%. This would effectively counteract the mutation (or even the deletion) of 1% of the functional genome. In this hypothetical situation, that single point mutation could create as much information as might be contained in 30 million nucleotide sites. In this manner, a few mega-beneficial mutations could theoretically counteract millions of deleterious point mutations.

Objection overruled:

The above scenario fails for four reasons:

a. The reductionist model of the genome is that the genome is basically a bag of genes and nucleotides, each gene or nucleotide acts largely in an additive manner. In such a model, essentially all information must be built up one tiny bit at a time – like building a pile of sand one grain at a time. This is even true in the special case of large DNA duplications. A duplicated region adds no new information until beneficial point mutations are somehow incorporated into it – one nucleotide at a time. The reductionist theoretician may give some lip-service to the importance of interactions and synergy - but in reality he knows that the only way to climb "Mount

Improbable" is through a very long series of infinitesimally tiny steps. We can ask ourselves, rationally: "What type of improvements might we hope for, via misspellings within a jet assembly manual?" Obviously any improvements, if they arose, would never involve large increments of improvement. At best they would involve very subtle refinements. At the heart of the Primary Axiom is slow incremental improvement. Under the Primary Axiom, we might safely say that a gene pool can only be filled up with information one tiny drop at a time.

b. In a genome having 3 billion units of information, the average beneficial mutation should only increase information by about one part in 3 billion. Yes, some beneficials will have more benefit than others, creating a natural distribution, but it is entirely unreasonable to believe that <u>any</u> beneficial point mutation could add as much information as let us say 30 million functional nucleotides.

   Some might object to this point as follows: *"There are certainly deleterious mutations which are lethal. In these cases a single point mutation can effectively negate 3 billion units worth of information. In fairness, shouldn't the reciprocal be true for beneficials – shouldn't the maximal beneficial mutation also be equal to 3 billion units of information?"* This line of thinking takes us back to the naïve view of mutation – a symmetrical bell-shaped curve. But we know that that view is universally rejected. Why?

   The extreme asymmetry of mutational effects has to do with the fact that one is trying to climb "Mount Improbable". Yes, it is conceivable that a mistake could cause you to stumble *up hill* – but only by a few feet. You will never stumble uphill by thousands of feet. However, a single error can easily cause you to fall *downward* very substantial distances. Indeed, while climbing Mount Improbable, you could easily plunge a very great distance - to your instant death. In the same way, if you are building a house of cards – failures are very easy, and are often very catastrophic - but you can only go *upward* one card at a time. In a very similar way, mutational changes are profoundly asymmetrical.

c. The concept of using a few mega-beneficial mutations to replace the information being lost at millions of other nucleotide sites is not rational, and leads to absurd conclusions. By this logic, just 100 mega-beneficial mutations, each of which might increase information by 1%, could replace the entire genome. One could delete all 3 billion bases within the genome and replace it with a genome consisting of just those 100 super-beneficial nucleotides. Indeed, if we could conceive of a mutation which actually doubled information (the mirror image of a lethal mutation), the entire rest of the genome could then be deleted. We would still have a fitness of one – based upon a genome of just one nucleotide! The mechanism of substituting a few mega-beneficials for millions of other information-bearing sites that are simultaneously being degraded by mutation - would result in an effective genome size that was continuously and rapidly shrinking. This is obviously impossible. It would be like trying to improve a book by subtracting 1000 letters for every new letter added.

d. Oft-cited examples of apparent "mega-beneficial mutations" are very misleading. For illustration, let us consider antibiotic resistance in bacteria, fur coat thickness in dogs, and homeo-box mutations in fruit fly.

   Chromosomal mutations within bacteria which confer antibiotic resistance appear to be mega-beneficial mutations. In the presence of antibiotic, the mutant strain lives, while all the other bacteria die. So fitness has not merely been doubled relative to the other bacteria – it has increased infinitely, going from zero to one!

   If you take a Samoyed (arctic) dog and put it in the Sonora desert it will die. A mutation to "hairless", will allow adaptation to the extreme heat – so the dog will live. Fitness has again increased from zero to one - an infinitely large increase!

   Certainly the two examples above are both "mega-beneficial mutations" in terms of adaptation to a specific environment. But they are both loss-of-function mutations that reduce net information within the genome. In terms of information content, they are both still *deleterious* mutations. Almost all examples of what appears to be mega-beneficial mutations, merely involve adaptation to a new environment. This is just a type of fine-tuning - it is not genome building. The dramatic nature of these types of changes is not because the organism has "advanced" in any real way, but is only because everything else has died! It is only relative to the dead competitors that the mutant is seen as "improved". These types of mutations do not increase information, or create more specified complexity, or create in any way a "higher form of life".

Very regrettably, evolutionists have treated two very different phenomenon - *adaptation to environments* and *evolution of higher life forms* - as if they were the same thing. We do not need to be geniuses to see that these are totally different issues. Adaptation can routinely be accomplished by loss of information or even developmental degeneration (loss of organs). However, development of higher life forms (representing more specified complexity), always requires a large increase in information.

There is a special class of mutations which can profoundly affect the development of an organism – e.g. mutations arising within what are called "homeo-box" genes. These mutations can cause gross re-arrangements of organs. For example, a single mutation can convert an insect's antennae into a leg, or can cause a fly to have four wings instead of two. These mutations certainly qualify as mega-mutations. Such dramatic changes in body form, arising from simple mutations, have greatly excited many evolutionists. This class of mutation has created a whole new field of speculation termed "EvoDevo" (evolutionary development). This type of mutation is widely assumed to provide the Primary Axiom with macro-beneficial mutations – even as might allow for evolutionary saltations (big jumps forward).

It is indeed conceivable that macro-alterations caused by homeo-box mutations might sometimes be beneficial. It is even conceivable that they might sometimes be beneficial in a substantial way. But how often would this realistically happen, and could such point mutations really counteract genome-wide degeneration?

In terms of a jet manual, a single misspelling might convert the command "*repeat loop 3 times*" to "*repeat loop 33 times*". Or a misspelling might convert the command "*attach assembly 21 into body part A*" into "*attach assembly 21 into body part Z*". These typographical errors could result in very profound changes in the shape of the airplane – but would they ever be beneficial? If they were beneficial, could they effectively offset the loss of information arising from millions of other misspellings – degrading all the other components of the plane?

We can acknowledge that, theoretically, homeo-box mutations might be modulated in useful ways. However, the actual examples given are in fact profoundly deleterious. The antennae-leg in the fly is actually just a monstrosity – it neither acts as an antennae nor a leg. The fly with the extra set of wings can not use them (they are not attached to muscles or nerves). Those useless appendages only interfere with the functioning of the normal pair of wings, and the mutant flies can barely fly. It should be obvious that some random changes within any instruction manual will produce gross aberrations within the finished product. But would this in any way support the idea that mega-beneficial mutations are happening? Would this suggest that one such macro-mutation could increase total genomic information by as much as 1% - equal to 30 million nucleotides? Would it suggest that one such a mutation could counteract the degenerative effects of millions of mutations happening throughout the rest of the genome? Obviously not!

In conclusion, as much as they might help prop up the Primary Axiom, mega-beneficial mutations can not honestly be invoked.

Objection #2 – Noise can be averaged out.

If a population is essentially infinite in size and is perfectly homogeneous, and if "noise" is both constant and uniform, and if there is unlimited time - then all noise effects will eventually be averaged out, and thus even near-neutrals might be subjected to selection. Under these conditions, it is conceivable that natural selection might eventually stop degeneration.

Objection over-ruled:

None of these basic requirements for eliminating noise, as listed above, are ever met in nature:

a. Population size is never infinite, and in the case of man, population size has only become substantial in the last several thousands years. Evolutionists assume in the case of man an effective evolutionary population size of only about 10,000. That small population would never have existed as a homogeneous gene pool, but would have only existed as isolated sub-populations – perhaps 100 tribes, each with about 100 individuals. Natural selection would have largely been limited to competition within each tribe. Under these conditions, there could be no significant noise averaging.

b. Noise is never uniform. In particular, environmental noise is highly inconsistent - both spatially and temporally. For a tribe within a given region, the most important source of non-genetic noise might result from climatic extremes, but for another tribe in another region it might be disease, and for another – predation. For many generations nutritional variation may be the main source of noise which is confounding selection, followed by cycles of disease or warfare. Under these conditions, there will be no significant noise averaging.

c. As fitness declines due to mutation accumulation, the genomic background itself will be changing. In reference to the selection for any given nucleotide, there will be progressively more and more noise from all the other segregating mutations which are accumulating. While some aspects of environmental noise will scale with fitness (thus diminishing proportionately as fitness declines), some aspects of environmental noise will not scale with fitness – e.g. noise due to natural disasters.

This latter type of noise, which does not diminish in concert with fitness decline, will grow progressively more disruptive to selection, as fitness declines. Noise which is continuously increasing can not be effectively neutralized by noise averaging.

d. When noise is high, selection becomes largely neutralized, resulting in a rapid and catastrophic accumulation of mutations. Given our low fertility and high mutation rate, there is little time for effective noise averaging to operate - prior to our extinction. Noise averaging, to the extent it is happening at all (see considerations above), requires a huge number of selection events before there can be significant averaging. Since the hypothetical evolutionary human population would have been small, the only way to have huge numbers of selection events would be to average over many generations. So noise averaging could only become effective over long periods of time. However, long before noise averaging might help to effectively refine the selection process, the human population would already be extinct (perhaps even before 1,000 generations). So fitness would reach zero long before any mutational equilibrium could be hoped for. Noise averaging, even if it is actually happening (see above considerations), does not appear to be sufficient to halt the degeneration process soon enough to stop error catastrophe and extinction.

Objection #3 – The failure of the Primary Axiom is not a serious challenge to evolutionary thought.

What does it matter if the Primary Axiom is fatally flawed and essentially falsified? The Primary Axiom is just one of numerous mechanisms of evolution, and so is not crucial to evolutionary theory. Evolutionary researchers just need some more time, and some more funding, to work out the few “minor kinks” in their various theories.

Objection overruled:

This position is *damage control* and is clearly false:

a. **There is only <u>one</u> evolutionary mechanism.** That mechanism is mutation/selection (the Primary Axiom). There is no viable alternative mechanism for the "spontaneous generation" of genomes. It is false to say that mutation/selection is only one of various mechanisms of evolution. Yes, there are several types of mutations, and yes, there are several types of selection, but there is still only one **basic evolutionary mechanism** (which is mutation/selection). The demise of the Primary Axiom leaves evolutionary theory without any viable mechanism. Without any naturalistic mechanism, evolution is not significantly different from any faith-based religion.

b. Darwin's only truly innovative concept was the idea that the *primary creative force* in nature might be natural selection. Yet he had no conception of genetics or mutation, and therefore had no conception of what was actually being "selected". So he was entirely ignorant of all the problems addressed in this book. His general view, that natural selection could explain all aspects of biology, was simply his vigorously advanced **philosophical** position. Not until much later did the neo-Darwinists synthesize genetics, mutation, and natural selection - creating the field of Population Genetics. Only then did Darwinism take on the appearance of real science. Ever since that time mutation/selection has been, and still remains, the singular lynch pin holding together all aspects of Darwinian thought.

c. Degeneration is the precise *antithesis* of evolutionary theory. Therefore the reality of Genetic Entropy is positively fatal to Darwinism. Many people are claiming that the concept of Intelligent Design can not be approached scientifically, but is only a matter of faith. However, it is obvious that in biology the "Null Hypothesis" of Intelligent Design is mutation/selection. We all know that to disprove a Null Hypothesis is to strongly support The Hypothesis. Therefore, any scientific evidence which demonstrates that mutation/selection can **not** create or preserve genomes, is sound scientific evidence **supporting** Intelligent Design.

Anzai, T. et al. 2003. Comparative sequencing of human and chimpanzee MHC class I regions unveils insertions/deletions as the major path to genomic divergence. PNAS 100: 7708-7713.

Bataillon, T. 2000. Estimation of spontaneous genome-wide mutation rate parameters: whither beneficial mutations? Heredity. 84:497-501.

Behe, M. 1996. Darwin's Black Box: Biochemical challenge to Evolution. The Free Press. NY, NY.

Bejerano, G. et al. 2004. Ultraconserved elements in the human genome. Science 304:1321-1325.

Bergman, J. 2004. Research on the deterioration of the genome and Darwinism: why mutations result in degeneration of the genome. Intelligent design Conference, Biola University. April 22-23. (in press).

Bernardes, A.T. 1996. Mutation load and the extinction of large populations. Physica ACTA 230:156-173.

Britten, R.J. 2002. Divergence between samples of chimpanzee and human DNA sequences is 5% counting indels. PNAS 99:13633-13635.

Carlsen, E. et al. 1992. Evidence for decreasing quality of semen during past 50 years. BMJ 305:609-613.

Chen, J. et al., 2004. Over 20% of human transcripts might form sense-antisense pairs. Nucleic Acid Research 32:4812-4820.

Crow, J.F. 1997. The high spontaneous mutation rate: is it a health risk? PNAS 94:8380-8386.

Crow, J.F. and M. Kimura. 1970. An Introduction to Population Genetics Theory. Harper and Row. NY, NY p. 249.

Dawkins, R. 1986. The Blind Watchmaker. Norton & Company, New York.

Demski, W. 1998. The design inference: eliminating chance through small probabilities. Cambridge University Press.

Dennis, C. 2002. The brave new world of RNA. Nature 418:122-124.

Elena, S.F. et al, 1998. Distribution of fitness effects caused by random insertion mutations in E. coli. Genetica 102/103:349-358.

Elena, S. F. and R.E. Lenski.1997. Test of synergistic interactions among deleterious mutations in bacteria. Nature 390:395-398.

Ellegren, H. 2000. Microsatellite mutations in the germline: implications for evolutionary inference. TIG 16:551-558.

Eyre-Walker, A. and P. Keightley. 1999. High genomic deleterious mutation rates in Hominids. Nature 397:344-347.

Felsenstein, J. 1974. The evolutionary advantage of recombination. Genetics 78: 737-756.

Flam, F. 1994. Hints of a language in junk DNA. Science 266: 1320.

Gabriel, S.B. et al. 2002. The structure of haplotype blocks in the human genome. Science 296:2225-2229.

Gardiner, K. 1995. Human genome organization. Current Opinion in Genetics and Development 5:315-322.

Gerrish, P.J. and R. Lenski. 1998. The fate of competing beneficial mutations in an asexual population. Genetica 102/103: 127-144.

Gibbs, W.W. 2003. The hidden genome. Scientific American. Dec.:108-113.

Gitt, W.. 1997. In the beginning was information. Literatur-Verbreitung Bielefeld, Germany.

Hakimi, M.A. 2002. A chromatin remodeling complex that loads cohesion onto human chromosomes. Nature 418:994-998.

Haldane, J.B.S. 1957. The cost of natural selection. J. Genetics 55:511-524.

Higgins, K. and M. Lynch. 2001. Metapopulation extinction caused by mutation accumulation. PNAS 98: 2928-2933.

Hirotsune S. et al. 2003. An expressed pseudogene regulates the mRNA stability of its homologous coding gene. Nature 423:91-96.

Hochedlinger, K. et al., 2004. Reprogramming of a melanoma genome by nuclear transplantation. Genes and Development 18: 1875-1885.

Holladay, P.M. and J.M. Watt. 2001. De-generation: an exponential decay curve in old testament genealogies. Evangelical Theological Society Papers, 2001. 52nd Natl. Conf., Nashville, TN Nov. 15-17, 2000.

Howell et al. 1996. Evolution of human mtDNA. How rapid does the human mitochondrial genome evolve? A. J. Hum. Genet. 59: 501-509.

Hoyle, F. 1999. Mathematics of Evolution. Acorn Enterprises, LLC, Memphis, TN.

Johnson, J.M. et al. 2005. Dark matter in the genome: evidence of widespread transcription detected by microarray tilling experiments. Trends in Genetics 21:93-102.

Karlin, S. 1998. Global dinucleotide signatures and analysis of genomic heterogeneity. Current Opinion in Microbiology. 1:598-610.

Kimura, M. 1968. Evolutionary rate at the molecular level. Nature 217: 624-626.

Kimura, M. and T. Ohta. 1971. Theoretical Aspects of Population Genetics. Princeton University Press, Princeton, NJ, pp 26-31, p 53.

Kimura, M. 1976. How genes evolve; a population geneticist's view. Ann. Genet.,19, no3, 153-168.

Kimura, M. 1979. Model of effective neutral mutations in which selective constraint is incorporated. PNAS 76:3440-3444.

Kimura, M. 1983. Neutral Theory of Molecular Evolution. Cambridge Univ. Press, NY, NY. (p.26, pp 30-31).

Kondrashov, A.S. 1995. Contamination of the genome by very slightly deleterious mutations: why have we not died 100 times over? J. Theor. Biol. 175:583-594.

Kondrashov, A.S. 2002. Direct Estimate of human per nucleotide mutation rates at 20 loci causing Mendelian diseases. Human Mutation 21:12-27.

Koop, B.F. and L. Hood. 1994. Striking sequence similarity over almost 100 kilobases of human and mouse T-cell receptor DNA.

Lee, J. 2003. Molecular biology: complicity of gene and pseudogene. Nature 423:26-28.

Lynch. M. et al. 1995. Mutational meltdown in sexual populations. Evolution 49 (6): 1067-1080.

Lynch, M., J. Conery, and R. Burger. 1995. Mutation accumulation and the extinction of small populations. Am. Nat. 146:489-518.

Manuelidis, L. 1990. View of interphase chromosomes. Science 250:1533-1540.

Mattick, J.S. 2001. Non-coding RNAs: the architects of eukaryotic complexity. EMBO reports 2:986-991.

Morrish, T.A. et al. 2002. DNA repair mediated by endonuclease-independent LINE-1 retrotransposition. Nature Genetics:31:159-165.

Morton, N.E., J.F. Crow, and H.J. Muller. 1956. An estimate of the mutational damage in man from data on consanguineous marriages. PNAS 42:855-863.

Muller, H.J. 1950. Our load of mutations. Amer. J Human Genetics 2:111-176.

Muller, H. J., 1964. The relation of recombination to mutational advance. Mutation Research 1:2-9.

Nachman, M.W. and S.L. Crowell. 2000. Estimate of the mutation rate per nucleotide in humans. Genetics 156:297-304.

Neel, J.V. et al. 1986. The rate with which spontaneous mutation alters the electrophoretic mobility of polypeptides. PNAS 83:389-393.

Ohno, S., T. Yomo. 1991. The gramatical rule for all DNA: junk and coding sequences. Electrophesis 12:103-108.

Paley, W. 1802. Natural theology: evidences of the existence and attributes of the Deity, collected from the appearances of nature.

Parson, T.J., et al. 1997. A high observed substitution rate in the human mitochondrial DNA control region. Nature Genetics 15:363-368.

Patterson, C. 1999. Evolution. Comstock Publishing Associates, Ithaca, NY.

Provine, W.B. 1971. The Origins of Theoretical Population Genetics. University of Chicago Press, Chicago. pp174-177.

ReMine, W. 1993. The Biotic Message. St. Paul Science, St. Paul, MN.

ReMine, W. 2005. Cost of Selection Theory. Technical Journal 19:113-125.

Sagan, C. "Life" *in* Macropedia, Encyclopedia Britannica, 1974 edition. pp 893-874.

Sandman, K. et al. 2000. Molecular components of the archaeal nucleosome. Biochimie 83: 277-281.

Schoen, D.J. et al. 1998. Deleterious mutation accumulation and the regeneration of genetic resources. PNAS95:394-399.

Shabalina, S.A. et al. 2001. Selective constraint in intergenic regions of human and mouse genomes. Trends in Genetics 17:373-376.

Shapiro, J.A., R.V. Sternberg. 2005. Why repetitive DNA is essential to genome function. Biol. Rev. 80:1-24.

Storz, G. 2002. An expanding universe of non-coding RNAs. Science 296:1260-1263.

Sutherland, G.R. and R.I. Richards. 1995. Simple tandem repeats and human disease. PNAS 92: 3636-3641.

Tachida, H. 1990. A population genetic model of selection that maintains specific trinucleotides at a specific location. J. Mol. Evol. 31:10-17.

Taft, R.J. and J.S. Mattick. 2003. Increasing biological complexity is positively correlated with the relative genome-wide expansion of non-protein-coding DNA sequences. Genome Biology 5(1):P1.

Tishkoff, S.A. and B.C. Verrelli. 2003. Patterns of human genetic diversity: implications for human evolutionary history and disease. Annual Review of Genomics and Human Genetics 4:293-340.

Trifonov E.N. 1989. Multiple codes of nucleotide sequences. Bull. of Mathematical Biology 51: 417-432.

Trifonov, E.N. 1997. Genetic sequences as product of compression by inclusive superposition of many codes. Molecular Biology 31 (4): 647-654.

Yelin, R. et al. 2003. Widespread occurrence of antisense transcription in the human genome. Nature biotechnology 21:379-386.

Vinogradov A.E., 2003. DNA helix: the importance of being GC-rich. Nucleic Acid Research 31:1838-1844.

Printed in the United States
200126BV00004B/1-132/A

9 781599 190020